PERSONAL TRANSPORT AND THE GREENHOUSE EFFECT

A Strategy for Sustainability

PETER HUGHES

EARTHSCAN

Earthscan Publications Ltd, London

Dedication
To M.I.H.

First published in the UK in 1993 by
Earthscan Publications Limited
120 Pentonville Road, London N1 9JN

A catalogue record for this book is available from the British Library

ISBN: 1 85383 169 7

Typeset by Saxon Graphics Ltd, Derby
Printed and bound by Biddles Limited, Guildford and Kings Lynn

Earthscan Publications Limited is an editorially independent subsidiary
of Kogan Page Limited and publishes in association with the International Institute for Environment and Development and the World Wide
Fund for Nature.

Contents

Preface

The research described in this book was made possible by a grant from the Science and Engineering Research Council, and carried out at the Open University's Energy and Environment Research Unit. The supervisor for the project was Dr Stephen Potter, and the SPACE model (see p60) was created using Microsoft Excel software.

I am indebted to the many individuals who provided information and guidance on aspects of the project, both in Britain and in Berkeley, California. They include Keith Buchan, Steven Cousins, Damien Dallemagne, Malcolm Fergusson, Phil Goodwin, David Howard, Jon Koomey, David Lowry, the late John Roberts, Marc Ross, Lee Schipper, Dan Sperling, Philip Steadman and John Whitelegg.

Peter Hughes

Note

Although the units **miles** and **gallons** are in still in use in Britain, the units **kilometres** and **litres** are used wherever possible in this book. This prevents any confusion that may arise between imperial and US gallons. The following are conversion factors for some of the quantities appearing frequently in the text.

1 kilometre = 0.6215 miles
1 litre = 0.220 gallons (= 0.264 US gallons)
Miles per gallon × litres per 100 kilometres = 282.5
1 megajoule (MJ) = 1,000,000 joules (J) = 0.278 kilowatt hours (kWh)
1 petajoule (PJ) = 1,000 terajoules (TJ)
$\qquad\qquad\quad$ = 1,000,000 gigajoules (GJ)
$\qquad\qquad\quad$ = 1,000,000,000 megajoules (MJ)

One day we walked down to Trafalgar Square. The tide was in, and the water reached nearly to the top of the wall on the northern side, below the National Gallery. We leant on the balustrade, looking at the water washing around Landseer's lions, wondering what Nelson would think of the view his statue was getting now...

She took my arm, and we started to walk westward. Halfway to the corner of the Square we paused at the sound of a motor. It seemed, improbably, to come from the south side. We waited while it drew closer. Presently, out from the Admiralty Arch swept a speedboat. It turned in a sharp arc and sped away down Whitehall, leaving the ripples of its wake slopping through the windows of august Governmental offices.

John Wyndham, *The Kraken Wakes*

1

Introduction

Since the affairs of men rest still uncertain,
Let's reason with the worst that may befall.

William Shakespeare, *Julius Caesar*

In the late 1980s, global warming grew in significance from a little-known scientific theory to an issue of deep concern for politicians, industrialists and the public alike. The realization that human activities could be shifting the world's climate into a new, warmer epoch by enhancing the atmosphere's natural 'greenhouse effect' is an issue that has grave implications for the entire world community.

Global warming is an alarming prospect, not only because of the likely impact that it will have on the Earth and the life that inhabits its surface, but also because of the overwhelming difficulties that are likely to be faced by world leaders in the search for solutions. Since the Industrial Revolution, the developed world has been increasingly locked into a lifestyle that demands unsustainable levels of energy and resource consumption. To make matters worse, developing countries are increasingly seeking the affluent, resource-intensive lifestyles to which the developed world has become accustomed, and in doing so are threatening to tip the ecological balance irreversibly towards catastrophe. The issue of 'sustainable development' is nowhere more critical than in the field of personal travel, which in many countries has become the fastest-growing contributor to 'greenhouse' emissions.

The environmental damage associated with personal mobility is not, of course, restricted to climate change alone. Air pollution in many European and North American cities has begun to reach chronic levels, with incidents of very poor air quality becoming steadily more common. Unlike the famous smogs of the 1950s, today's toxic fumes are produced not by coal-burning in houses and factories but by the use of automobiles. In addition to urban air pollution, the manufacture of vehicles, principally private cars, imposes a heavy toll on the natural environment.

Meanwhile the use of cars, particularly in urban areas, has created landscapes of concrete and tarmac that create a sense of alienation for the people who live there.

The inter-urban road network has provided fertile ground for property developers, as lax planning regulations have allowed them to create massive retail and leisure parks on greenfield sites flanking motorways and major roads. By creating facilities which can be reached only by car, developers have not only inflicted commercial hardship on traditional town and city centres, but contributed to society's increasing dependence on the automobile.[1]

Mobility is not something that we can give up easily. But in view of the ever-increasing demand for cars, roadspace and fossil fuel, we have no choice but to examine what exactly it is that mobility provides. Almost without exception it is the *access* – to people, goods and services – which we value so highly, rather than the travel itself. As Dr Phil Goodwin and his colleagues at Oxford University have written, in *Transport: The New Realism* (1991):

> [Transport] is unlike many other fundamental human activities in that in most cases movement is a means to an end, not an end in itself. There are exceptions, such as a pleasure cruise, or walking the dog, but for most day to day trips we do not really want to travel at all – we want to participate in some activity in a different place and transport is simply something we have to do to enable this.

Obvious though this observation may seem, it holds the key to resolving the conflict between personal travel and its environmental impacts. By focusing attention on the actual benefits that our mobility-intensive lifestyles deliver, we can begin to look for alternative, less damaging ways of fulfilling society's basic need for access and communication. While traditional transport policies have concentrated on the movement of vehicles, planners are now turning towards the movement of *people* as their key aim. For example, cars have historically tended to be given greater priority than bicycles in transport planning, because of their greater size and speed. However, one may question why a person driving to an out-of-town supermarket should be given any greater priority than a person cycling 2 kilometres to the local shops: both individuals are gaining the same benefit from their trips, even though the economic and environmental costs of the two journeys are so widely different. Similarly, there is a strong case for giving priority to buses, which generally carry many times more people than private cars.

Technological advances are making it increasingly possible to replace journeys with digital communications. 'Telecommuting' or 'teleworking', explored later in the book, is becoming a viable option for many office-bound workers, offering them the possibility of avoiding travel altogether on certain days of the week. But teleworking will only ever be a partial solution, not least because it is *leisure* travel, rather than journeys to and from work, that is currently growing most rapidly.

One might be forgiven for thinking that technology holds all the answers to the travel-versus-environment conflict, and that society can simply employ more scientists and engineers to develop environment-friendly ways of maintaining the mobility that it has come to enjoy. But evidence to date strongly indicates that technological advances can only be part of the solution, and significant lifestyle changes will be an essential part of any strategy for sustainable transport. For example, even if it were possible to replace the entire car fleet of Europe and North America with 'zero-emissions vehicles' (an unlikely proposition), society would still be faced with the massive environmental damage that arises from the manufacture and disposal of cars, and from the construction of roads and carparks on which to operate them.

National governments in Europe and North America have been keen to play down the environmental impacts of car use. Anxious to support their national car industries, yet aware of a growing level of environmental concern among their electorate, governments have been happy to support the myth of the environment-friendly car. Unleaded petrol, catalytic converters, airbags and vehicle component recycling have all contributed to the illusion of green and friendly motoring. But none of these measures can do more than scratch the surface of the problem. The real issue at stake is society's deep-seated dependence upon the car, and the associated issues of oil consumption and greenhouse gas emissions arising from it.

In 1989 the German car manufacturer Volkswagen-Audi ran a full-page newspaper advertisement in Britain's daily press outlining the measures that the company has taken to reduce toxic exhaust emissions from its cars, through the use of catalytic converters. The advert explained how poisonous gases produced by the engine enter the catalytic converter and undergo chemical reactions, which turn them into 'harmless carbon dioxide, nitrogen and water.' If carbon dioxide – the most prolific greenhouse gas produced by human activities – really is harmless, as Volkswagen apparently believes, then there is little point in reading any further. But a consensus of opinion among the scientific community would suggest otherwise.[2]

It is, of course, easy to blame car manufacturers for the environmental damage wreaked by mass mobility. After all, one new car is brought into the world every second. But in reality it is society that is responsible for the problem, and which must therefore work to find solutions. Every minute, more and more people in the developed world acquire cars. Every day, we consume products that are brought to us via an energy-intensive network of freight distribution.[3] And every year, many of us are more than happy to take foreign holidays by air, often consuming as much petroleum in eight hours as we would do in a year travelling by car.

This book proposes a realistic strategy by which personal travel can be reconciled with the need to stabilize the atmosphere and halt the build-up of carbon dioxide, the principal greenhouse gas. It is based not on arbitrary targets for reducing emissions, but on the scientific community's best estimate of the reduction in emissions that will be necessary in order to stabilize the atmosphere. It takes a realistic view of the kind of measures that society is likely to accept as part of a strategy for reducing greenhouse emissions. It would, after all, be easy, but entirely unproductive, to recommend that the use of cars be banned overnight, and to suggest that the use of public transport be made compulsory. The environmental benefits of such a policy would, of course, be enormous – but the democratic process would not tolerate such a draconian response. Policies to protect the environment can only work if they have the support of the electorate. Well-meaning environmental organizations have been known to sink without trace by alienating the public on whose support they depend.

In 1987, the discovery of the Antarctic ozone hole by British scientists prompted a global response on an unprecedented scale. The shock of the discovery, compounded by the grave human and ecological consequences of ozone depletion, produced an effective and immediate response from the global community. The Montreal Protocol, signed in 1987 and last revised in 1992, represents a commitment by governments world wide to cooperation in the elimination of ozone-destroying gases. Subsequent measurements of stratospheric ozone levels have vindicated the sense of urgency with which this agreement was signed.

Many have pointed to the Montreal Protocol as an analogy of the kind of global consensus that will need to be secured in order to avert a greenhouse catastrophe. At the time of writing, most governments have acknowledged the likelihood of climatic change through emissions of greenhouse gases, but have made no more than a minor response. The purpose of this book is to propose ways in which governments and their electorates can contribute to a coordinated, international programme for averting global warming.

It first describes the current state of knowledge in the field of climatology and the likely impact of 'anthropogenic' – human generated – greenhouse gas emissions. Much of this review is based on the work of the Intergovernmental Panel on Climate Change (IPCC) scientific group, a body set up by the United Nations to establish best estimates of the nature of the greenhouse problem. The book then examines how, and to what extent, personal travel contributes to the greenhouse effect, with an examination of each of the different forms of transport in turn.

The book then turns to carbon dioxide (CO_2), the most abundant greenhouse gas produced by human activities, and the main focus of this work. In order to estimate how emissions of this gas might change under different possible futures, a computer model called SPACE (Scenario Projections of Aggregate Carbon Emissions), has been developed. As a first step, the SPACE model is used to predict how transport's emissions are likely to change in a 'do nothing' world, based on a 'business as usual' scenario, which assumes a continuation of current policies and no intervention from central government.

A second scenario is then constructed to address the question of whether, and to what extent, technological solutions might be found to the transport-greenhouse problem. The importance of this question is self-evident: if 'technical fixes' can be used to eliminate CO_2 emissions from personal travel, then we may consider the problem solved. If, on the other hand, technology does not hold the whole answer, then it will be necessary to go beyond technical measures in order to find an environmentally sustainable solution.

A third and final scenario examines what could be done if all the available policy measures were to be called upon at once. If all the stops were pulled out, what reduction in CO_2 emissions could we expect to achieve? The answer to this question is of immense relevance to all developed countries – not to mention the developing world, where the disturbing spectre of mass car ownership is beginning to materialize.[4]

Although the three scenarios are modelled on personal travel in Great Britain, the strategy developed in this book could equally be applied to most European countries. Britain's problems are fairly typical: on the one hand, more and more people are acquiring and using cars, and travelling further every year; while on the other there is pressure to meet a national target of stabilizing CO_2 emissions by the year 2000. Care should be taken, however, in extrapolating the results to countries such as the United States, where the nature of personal travel is considerably different from that in Europe.

Politicians in developed countries are faced with many critical issues, including domestic economics, international stability, Third World debt and environmental protection. Added to this list is the threat of irreversible climate change arising from the build-up of greenhouse gases in the atmosphere. Although less tangible than most of the day-to-day concerns affecting politicians, the threat of global warming is in many ways the most serious of all. This book aims to steer a course through the technical and political obstacles, and sets out the steps that could be taken to lessen the conflict between personal mobility and long-term environmental security.

Notes

1. A catalogue of these and other environmental impacts of transport can be found in Whitelegg (1993), whose rigorous examination of 'sustainable' transport challenges many of the assumptions upon which current transport policy is based.
2. A second advertisement, published later in the year by Volkswagen-Audi, added insult to injury by describing carbon dioxide as 'the stuff that makes fizzy drinks fizzy'.
3. Whitelegg (1993) estimates that each person in Britain consumes the equivalent of 60 tonne-kilometres of freight per week.
4. See, for example, Meyers, 1988.

2

Global Warming and Climatic Change

> Naturally occurring greenhouse gases keep the Earth warm
> enough to be habitable. By increasing their concentrations,
> and by adding new greenhouse gases like chlorofluorocar-
> bons, humankind is capable of raising the global temperature.
>
> Intergovernmental Panel on Climate Change

In the late 1980s global warming grew in prominence from a little-known climatic theory to a major concern on the public and political agenda. The possibility that anthropogenic emissions could be altering the behaviour of the Earth's climate system gained credibility following a series of climatic anomalies world wide, including floods, droughts and hurricanes of extraordinary ferocity. Six of the seven warmest years ever recorded fell in the 1980s, and 1990 was the warmest year since records began.

The 'greenhouse effect' is the popular term for the principle underlying global warming. The Earth's surface is heated by radiation from the Sun, and most of the energy is then radiated back into space at infrared (IR) wavelengths. However, trace gases present in small quantities in the lower atmosphere reabsorb a small amount of this outgoing radiation, warming the atmosphere and the surface of the Earth. Without this natural greenhouse effect, the average temperature at the Earth's surface would be an inhospitable −19°C, compared with the actual average of 15°C.

In fact the analogy of a greenhouse is not strictly correct. The warming effect of a garden greenhouse is produced by a thin layer of glass that retains the warm air inside. In the atmosphere, greenhouse gases are present not as a thin layer but spread throughout the atmosphere in tiny concentrations, and they warm the atmosphere by absorbing radiation rather than by restricting the escape of warmed air.

Under equilibrium conditions, the amount of energy from the Sun entering the atmosphere is exactly balanced by the energy radiated back into space. The average temperature of the atmosphere settles at a level in which this balance is established. Any additional factor that disrupts the balance by increasing the absorption of heat in the atmosphere is known as a *radiative forcing agent*. The atmosphere tends towards conditions in which incoming and outgoing radiation are in equilibrium. Therefore if radiative forcing is increased as a result of rising levels of greenhouse gases, the result will be an increase in atmospheric temperature.

The most abundant of the 'greenhouse gases' is *water vapour*, which is present in large quantities in the atmosphere. In second place is *carbon dioxide* (CO_2), which is circulated in the biosphere as part of the natural 'carbon cycle'. Carbon is exchanged between the atmosphere, the oceans and the land via a complex network of biological and chemical processes. But since the Industrial Revolution, human activities have created an imbalance in the carbon cycle and increased the concentration of CO_2 in the atmosphere. This has taken place in two ways.

Firstly, some of the 'sinks' for CO_2, in the form of forests and other forms of vegetation, have been removed in the process of industrialization. Forests form part of the CO_2 recycling process, and large quantities of carbon are 'stored' in the plants that they contain. By destroying these plants, and particularly trees, humans have released carbon to the atmosphere in the form of CO_2. An area of Amazonian rainforest the size of Britain is currently disappearing each year. At this rate, the world's rainforests will be completely destroyed by the year 2010.

Secondly, the large-scale consumption of fossil fuels – coal, oil and gas – since the Industrial Revolution has released large amounts of CO_2 directly into the atmosphere, as CO_2 and water vapour are the principal combustion products when hydrocarbons are burned. Fossil fuels have been put to a wide variety of uses, including heating, cooking, lighting, manufacturing, and transporting goods and people. It is the last of these categories that forms the focus of this book.

Besides CO_2 and water vapour, other greenhouse gases have more recently begun to increase in concentration. In the hundred years preceding 1980, CO_2 accounted for 66 per cent of radiative forcing. In the 1980s the contribution of CO_2 emissions to global warming had fallen to 49 per cent of total radiative forcing, with the other 51 per cent due to other anthropogenic gases (Lashof and Tirpak, 1989).

Nitrous oxide (N_2O) is some 250 times more powerful as a radiative forcing agent than CO_2 per molecule, and results primarily from the combustion of fossil fuels and biomass, as well as the use of fertilizers. *Methane* (CH_4) is 25 times more powerful than CO_2 and is produced when organic matter decomposes anaerobically. Major sources include leakage from gas pipelines, emissions from rice paddies, and fermentation in the digestive systems of livestock.

Chlorofluorocarbons (CFCs) and *halons* are a relatively recent invention, valued for their inert properties. They are up to 20,000 times more powerful than CO_2 as greenhouse gases. These gases have been used as aerosol propellants, refrigerants and foaming gases since the 1930s. The discovery in the late 1980s that CFCs are rapidly depleting the stratospheric ozone layer prompted a considerable reduction in their use, as this chapter will later explain. The atmospheric concentration of these gases had previously been rising at a growth rate of 5 per cent per year. Substitutes for CFCs are being developed by the chemical industries, in the form of hydrochlorofluorocarbons (HCFCs) and hydrofluorocarbons (HFCs). These are generally less damaging in terms of ozone depletion and radiative forcing, but are nevertheless significant greenhouse gases (Fisher et al, 1990).[1]

Finally, ground-level *ozone* (O_3) – not to be confused with the ozone layer in the upper atmosphere (stratosphere) – is also thought to have a radiative forcing effect. It is formed by the reaction between nitrogen oxides (NOx) and unburnt hydrocarbons in the presence of sunlight, as well as in some electronic machinery. Ozone at ground level is also known for its adverse effects on human health. The exact contribution of tropospheric ozone to global warming has not yet been established (Watson et al, 1990).

Table 2.1 lists the principal greenhouse gases together with their current levels, growth rates and contributions to global warming. It can be seen that almost three-quarters of greenhouse emissions from human activities are in the form of CO_2. The figures also show that the relative *lifetimes* of the different greenhouse gases vary considerably. The enormous global warming potential of CFCs is illustrated by the substantial contribution that they make to global warming, despite their relatively tiny concentrations in the atmosphere.

Table 2.1 Greenhouse gases produced by human activities

	Carbon dioxide	Methane	CFC11	CFC12	Nitrous oxide
Atmospheric concentration (ppm by volume)					
Pre-industrial (1750-1800)	280	0.8	0	0	0.288
Present-day (1990)	355	1.72	0.0003	0.0005	0.310
Current annual rate of change	0.5%	0.6–0.75%	4%	4%	0.2–0.3%
Lifetime (years)	120	10.5	55	116	132
Current contribution to global warming	72%	10%	13%		5%

Note ppm = parts per million
Source Department of the Environment, 1993

Table 2.2 lists the relative emissions of the 20 nations which are the greatest contributors to greenhouse emissions, in terms of 'carbon equivalent' – a useful measure representing the combined effect of all greenhouse gas emissions, expressed as an amount of CO_2 that would have the same overall effect as all of the greenhouse gas emissions combined. In 1991, total global emissions of CO_2 from man-made sources were in the region of 27,000 million tonnes, most of which was produced by the industrialized countries. Taken together, the countries of the OECD and former USSR account for two-thirds of the CO_2 added to the atmosphere each year. The largest single contributor was the USA, which produced around a quarter of the total. The former USSR was responsible for around 17 per cent, while Western Europe produced 15 per cent of total CO_2 (MacKenzie, 1990).[2]

Table 2.2 The 20 largest contributors to greenhouse forcing in 1988

Country	% contribution to global warming	'Greenhouse' emissions per capita (USA=100)
United States	17.1	100
Australia	1.1	93
Canada	1.6	89
Myanmar (Burma)	2.1	74
USSR	13.5	67
Western Germany	2.7	63
United Kingdom	2.4	59
Brazil	5.7	56
Poland	1.4	54
Colombia	1.1	54
Italy	1.8	44
Japan	3.6	43
France	1.7	43
Spain	1.1	39
Thailand	1.2	31
Mexico	1.6	28
Indonesia	1.2	15
China	8.1	11
India	4.6	7
World	**100.0**	**28**

Source Hammond et al, 1991

Historical trends

Historical concentrations of CO_2 can be deduced through the analysis of ice cores. By drilling into ancient glaciers and analysing the air bubbles frozen within them, an extensive history of global CO_2 levels can be obtained. The longest ice core drilled to date is over 3,000 metres long, and the information contained within it goes back over 200,000 years (Radford, 1992). Present-day concentrations of CO_2 can be measured using more sophisticated optical absorption techniques.

Meanwhile historical temperatures can be deduced from geological tests, involving analysis of the isotopes contained in ancient ice formations and in sediments from oceans and lakes. On a timescale of millions of years, greenhouse warming, primarily by CO_2, is believed to have decreased.

But superimposed upon this trend have been shorter-term fluctuations in the atmospheric CO_2 concentration, on timescales of thousands of years rather than millions, associated with major climatic events such as ice ages. It is believed that changes in the Earth's orbit around the Sun may be responsible for these periodic transitions between glacial and interglacial conditions (Gribbin, 1989).

The most recent glaciation, when the atmospheric CO_2 concentration fell to around 210 parts per million (ppm), began around 100,000 years ago and ended around 10,000 years ago. After this the atmospheric CO_2 level rose to a steady 280 ppm, the concentration found in air samples that have been trapped in polar ice for several thousand years. CO_2 remained at 280 ppm until around 1860, when the onset of industrialization began to increase sharply the atmospheric concentration. Since then, the widespread combustion of fossil fuels and the removal of forests has caused the concentration to rise by approximately 0.5 per cent per annum to its present level of around 355 ppm (see Figure 2.1). In little more than a hundred years, the atmospheric CO_2 concentration has risen by 25 per cent.

Climatic change

In order to predict the likely effects of a warmer atmosphere, *general circulation models* (GCMs), running on powerful computers, are employed. GCMs are adaptations of models used in weather forecasting, which divide the Earth's surface into grid squares typically 300 km across and solve the climatic equations at regular time intervals. To date there has been a broad consensus of opinion amongst modellers about the nature of global warming. It is predicted by the IPCC that an effective doubling in CO_2 concentration will raise the average global temperature by between 1.5 and 4.5°C. ('Effective doubling' refers to an increase in the concentration of *all* greenhouse gases equivalent in effect to a doubling of CO_2 alone.) At present rates of emission, effective doubling will take place around 2030 (Leggett, 1990).

Recent measurements of global temperature have demonstrated an unmistakable warming trend. Several years in the 1980s were the hottest ever recorded. But these observations, despite their serious implications, cannot be conclusively linked to a long-term warming trend. The global temperature record contains a significant element of 'noise', or short-period variability, from which sustained trends are hard to extract.

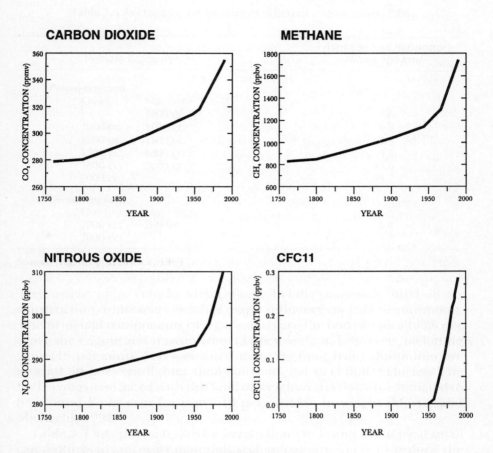

Notes ppmv = parts per million (volume)   ppbv = parts per billion (volume)

Source Houghton et al, 1990

Source Houghton et al, 1990

Figure 2.1 Atmospheric concentration of greenhouse gases since 1750

In addition to the 'noise' problem, there are a number of alternative explanations for the observed warming trend. One theory links global warming with the low occurrence of volcanic activity in recent years. Fewer volcanoes mean less dust in the atmosphere, and hence less sunlight blocked out. The net result is thought to be a warmer atmosphere. Another explanation involves variations in solar activity, which are known to take place over an 11-year cycle. Finally, the temperature rise could simply be a continuation of the warming associated with the end of the European 'mini ice age' that occurred some 300 years ago – though the rapid rate of warming does suggest that some additional influence is involved. In particular, it should be remembered that the rate of warming forecast by the IPCC is unprecedented in world history. Following the last glaciation, the temperature rose at an average rate of around 1°C per 500 years. By contrast, the IPCC 'best guess' prediction represents a warming rate of 1°C every 30 years.

A rapid warming of the Earth's atmosphere would have a variety of adverse impacts. Among the more serious consequences of a warmer climate is the likelihood that global sea level will rise. Most of the rise would be due to thermal expansion of the oceans, with some resulting from the melting of land-borne glaciers. The IPCC estimates that sea levels will rise by between 10 and 30 cm by the year 2030, with a 'best guess' of 20 cm. By 2100 it is predicted that sea levels will have risen by 30 to 100 cm, with a best guess of 65 cm. Given that a third of the world's population inhabits coastal regions, the consequences of losing these areas to the sea are very grave indeed. Cities particularly at risk from rising sea levels include Los Angeles, Miami, London, Amsterdam, Bombay, Shanghai, Hong Kong and Sydney. Many coastal cities would incur severe difficulties as a result of even a modest increase in sea level such as the 65 cm predicted by the IPCC.

Another likely consequence of global warming would be a change in rainfall patterns, which would be likely to endanger millions of lives through starvation. Large numbers of 'environmental refugees', combined with a steadily rising global population, would be likely to have a destabilizing effect on global politics. As Stephen Schneider of the National Centre for Atmospheric Research in Boulder, Colorado, puts it:

> We have a world of over 5 billion people, with people tightly locked into national boundaries. Migration is not as easy as it used to be; you can't move away from a bad climate to a good one any more, and we have maybe a billion people living in marginal conditions. These are very difficult problems with or without climate change. If you add rapidly changing climate, this puts another stress right on top of an already stressed world struggling for economic development.

Viewpoint '89 (1989)

Concern has also been expressed in connection with global food production. Much of the agriculture upon which the world's population depends takes place in 'marginal' conditions, where a small change in climate could virtually wipe out agricultural production. Recent droughts in the corn belts of the USA and the former USSR, as well as in less developed regions such as north-east Africa, have demonstrated the fragile nature of extensive agriculture – and also led to speculation as to whether droughts will become more commonplace as global warming takes hold. Regions identified by the IPCC as being particularly at risk include Northern and Southern Africa, India, the south-west USA, Central America and the Mediterranean. In other areas, particularly those at higher latitudes, it is predicted that global warming will lead to increased rainfall, which itself could cause serious disruption to agriculture.

It is probable that any increase in radiative forcing will be associated with a growth in the number, and severity, of climatic adversities. There are indications that the number of droughts, floods, storms and so on, when measured as a total over each of the last three decades, has risen substantially (Timberlake, 1988). The year 1988, for example, was a year of great droughts in the corn belts of the USA and the former USSR, and the catastrophic hurricanes Joan and Gilbert which severely disrupted the ecology of Nicaragua and the Caribbean. Speculation regarding the effects of global warming upon the polar icecaps was heightened in 1987 when an iceberg 150 km long broke away from Antarctica, then drifted northwards and broke up. In September 1989, ice in a region to the north of Greenland was reported to have thinned from 6.7 to 4.5 m between 1976 and 1987. Similar effects were observed in the Antarctic, where average air temperatures rose by 1.1°C between 1982 and 1986 (Highfield, 1989).

Other unusual weather phenomena have been observed but, by their complex and unpredictable nature, provide little contribution to the evidence for global warming. Although one of the main predictions made by climate modellers is an increase in the severity and frequency of climatic adversities, they cannot say for sure whether a particular drought or flood is the direct result of global warming or simply a 'natural' event. Nor is it possible to predict with any confidence where and when the next climatic anomaly might take place. For example, 3,000 people lost their lives in the storms and tidal surges that swamped Bangladesh in September 1988. This type of event is consistent with the onset of global warming – yet it is impossible to say whether this particular disaster was 'caused' by the greenhouse effect.

Equally important is how the weather patterns of individual continents would be affected by global warming. For example, there is a possibility of the 'Mediterranean' climate spreading northwards across Europe, bringing drier, warmer conditions to many countries. But there is also speculation that the temperate Gulfstream which warms northern Europe during the winter could be affected, leading to colder winters.

Many natural habitats are threatened by the prospect of a changing climate. For many species, the rapid *rate* of temperature change poses the greatest threat. Dobson et al (1989) have suggested that the effect of a one-degree rise in temperature is equivalent to a change in latitude of up to 160 km, and the IPCC 'best guess' indicates that such a change could take place in a period of just 30 years. Many species of flora and fauna would have to migrate towards the Poles in order to find favourable conditions, and it is likely that the necessary speed of migration would be too great for some plant varieties. As a result, many habitats would be lost altogether. Meanwhile some species, such as the migrating fish of the Atlantic, could be threatened if the natural signals which prompt their reproductive activity are disrupted.

Climate change would have a significant impact on human health, in both developed and developing countries. The effects of global warming upon food production clearly have serious implications for nutrition, while the spread of communicable diseases, particularly in the Third World, is likely to increase as a result of higher temperatures. In the developed world, other possible health hazards include temperature-related disorders and a deterioration in mental health resulting from experience of environmental catastrophes (Haines, 1990).

Finally, it should be noted that the effects of global warming may be amplified by a number of possible feedback mechanisms. Only some of these feedbacks are included in GCMs; for others, it has not yet even been established whether the effect is positive or negative in nature. For a fuller discussion of these phenomena see Schimel (1990) and Woodwell (1990). Perhaps the greatest uncertainty in the global warming debate concerns the way in which the Earth's natural systems might react to a CO_2-enriched atmosphere. Of particular concern are the microscopic life-forms that inhabit the oceans: their photosynthetic activity might be expected to increase in a warmer world, absorbing more CO_2 from the atmosphere and increasing the build-up of this gas; alternatively, they may be wiped out by any change in ocean temperature, their disappearance contributing further to the rising concentration of CO_2 in the atmosphere.

Another feedback mechanism is related to the behaviour of clouds. It is possible that there would be increased cloud cover in a warmer world, leading to a reduction in the amount of sunlight reaching the Earth's surface and warming the atmosphere. Alternatively, melting icecaps in polar regions may release large amounts of methane from the thawing subsoil, contributing further to radiative forcing. The loss of ice might also lead to increased absorption of solar heat due to reduced surface albedo (whiteness).

Future greenhouse emissions

If emissions of CO_2 continue at present rates, the concentration of this gas in the atmosphere may be expected to reach almost 500 ppm by the year 2100 – 80 per cent higher than the pre-industrial level. Of crucial importance is the observation that stabilizing emissions is *not* equivalent to stabilizing atmospheric concentration. In order to halt the growth of CO_2 in the atmosphere, significant reductions in emissions are required. The IPCC has estimated that emissions of CO_2 from industrialized countries will need to be reduced by at least 60 per cent in order to stabilize the atmospheric concentration of this gas (see Table 2.3).

Table 2.3 Cuts in emissions needed for a stabilization of
greenhouse gases at present atmospheric levels

Greenhouse gas	IPCC estimated cut (%)
CO_2	at least 60
CH_4	15–20
N_2O	70–80
CFC11	70–75
CFC12	75–85
HCFC22	40–50

Source Houghton et al, 1990

Over 150 nations have now signed the United Nations Framework Convention on Climate Change, which commits them to returning emissions of CO_2 to 1990 levels by the year 2000. But according to F Pearce (1993), 'the Climate Change Convention will probably not stabilise global emissions and certainly will not halt the growth of CO_2 in the atmosphere'.

The response of individual countries to the objective of controlling CO_2 emissions is largely a matter of national policy. Within the European Community, the situation is varied: for example, Germany has opted for a 25 per cent cut by 2005, while Denmark, Italy and the Netherlands are also aiming to reduce net emissions. Britain has resolutely resisted calls for the introduction of a Europe-wide 'carbon tax' on energy and emissions, preferring instead to find its own ways of controlling CO_2 emissions.

However, the objective of 'stabilizing' CO_2 emissions at 1990 levels is, from a scientific point of view, an unacceptably modest target, and one that many developed countries are likely to attain even without any policy changes. The decline of heavy industry, coupled with a small shift away from fossil fuels to nuclear power, has led to a gradual decline in CO_2 emissions from many countries over the last 20 years (ibid).

Among industrial nations, it is possible to identify three broad categories of response to the threat of global warming:

■ *No response*: in the light of scientific uncertainty over the consequences of global warming, no national target for emission controls is established.
■ *Stabilize emissions*: a target is set for holding emissions down to present-day levels, typically by 2000 or 2005.
■ *Reduce emissions*: in order to approach the goal of stabilizing the *atmospheric concentration* of greenhouse gases, reduction strategies are established. Typically cuts of around 20 per cent by 2005 are envisaged, while greater reductions in emissions may be anticipated in the period following 2005.

The USA and former USSR have not agreed to set a target for controlling CO_2 emissions, but are nevertheless giving attention to areas in which reductions might be achieved. The USA's response has been typified by the so-called 'no regrets' approach, involving measures that are known to be beneficial in terms of reducing greenhouse emissions but which impose no additional costs upon society. The rationale underlying 'no regrets' is that if a particular policy measure is later deemed to have been unnecessary (perhaps if global warming fails to materialize on the scale presently anticipated) there will be no overall loss to society from having adopted that policy.

The value of 'no regrets', or the 'precautionary principle' is succinctly highlighted by R K Pachauri, director of the Tata Energy Research Institute. 'If we wait until conclusive evidence comes along, then the form of that evidence will be so expensive, in terms of the effect of life on this planet, that we will be able to do nothing about it' (*Viewpoint '89*, 1989).

Ozone depletion and the Montreal Protocol

In 1987, public alarm was raised by the discovery of a large 'hole' in the stratospheric ozone layer above Antarctica. Concern focused primarily on the health hazards of ultraviolet radiation, from which the ozone layer shields the Earth. Ultraviolet is associated with a variety of diseases including skin cancers and cataracts. However, public concern over the ozone-depleting effect of CFC emissions tended to eclipse the fact that CFCs are also a major contributor to global warming.

The discovery of the ozone hole prompted the first ever international treaty for environmental protection, the Montreal Protocol. The outcome of the negotiations, completed in September 1987, was a broad agreement to reduce the production and emission of five CFCs and three halons.

Renewed urgency was given to the issue of ozone depletion in April 1991, when observations from NASA's satellite-borne Total Ozone Mapping Spectrometer (TOMS) revealed that northern mid-latitudes were losing ozone at a greater rate than was previously thought. Since its original signing, the terms of the Montreal Protocol have been updated on two occasions,[3] with more stringent targets introduced for the phase-out of ozone-destroying gases. In addition, India and China have agreed to join the Protocol's signatories. The production of CFCs is now to be eliminated altogether by the end of 1995, and halons by the end of 1993, in the 90 countries concerned. The European Community has since agreed to bring forward the target date for the elimination of CFCs to the end of 1994 (Department of the Environment, 1993).

Whilst international difficulties are likely to arise in the attempt to secure CO_2 reduction targets, it appears likely that a virtual elimination of CFCs and halons will be achieved in the near future, leading to an eventual reduction in the atmospheric concentration of these gases. But CFCs have a relatively long lifetime, and it takes over 100 years for them to dissipate once they have reached the upper atmosphere (see Table 2.1). It should also be noted that some of the gases that are used in place of CFCs also have a significant radiative forcing effect, and from a global warming perspective should not be regarded as entirely benign.

Summary

A consensus of scientific opinion, embodied by the United Nations Intergovernmental Panel on Climate Change, has agreed that global warming will be a likely consequence of the build-up of anthropogenic greenhouse gases – particularly CO_2 – in the atmosphere. But there remains considerable uncertainty over the likely scale of the warming effect and its influence upon global climate and the world's ecosystems. However, it seems likely that global warming would create social and economic problems on a massive and probably unprecedented scale.

To date, there has been a wide variation between the responses of different countries. The most common strategy among industrialized nations, as embodied in the UN Framework Convention, is to stabilize CO_2 emissions at present levels, as a first step towards more stringent reductions. But the IPCC has indicated that a reduction of at least 60 per cent will be necessary in order to stabilize the atmospheric concentration of this gas.

Preliminary actions have been taken in order to curb the emission of CO_2 and other greenhouse gases into the atmosphere. In particular, the international convention for phasing out CFCs will have a beneficial effect in terms of global warming. But measures taken so far are likely to prove insufficient to avert further warming. If the global warming theory proves to be correct, it will be necessary to secure substantially greater reductions in emissions than those currently planned.

Notes

1. Significantly, evidence has recently come to light suggesting that the global warming effect of CFCs may be cancelled out to some extent by the cooling that is thought to take place as a result of the thinning of the ozone layer (World Meteorological Organisation, 1991).
2. Emissions from the UK amounted to around 583 million tonnes, around 2 per cent of the world total (Department of the Environment, 1993).
3. June 1990 and November 1992.

3

Travel, Energy Use and 'Greenhouse' Emissions

> In the United Kingdom, as in most other industrialized countries, combustion of fossil fuels for many purposes including transport is the major source of emissions which pollute the air.
>
> Len Watkins, Transport Research Laboratory

In order to quantify emissions of greenhouse gases from transport, and more particularly from personal travel, it is necessary first to define the transport system and its boundaries. Estimates of energy consumption and greenhouse emissions in transport are often restricted to the day-to-day running of vehicles: energy used in non-operational activities, such as the manufacture and maintenance of vehicles and infrastructure, tends to be overlooked – not least because of the difficulties involved in measuring it.

Howard (1990) has examined the relative energy demands of different aspects of transport, and estimated that if a 'wide view' is adopted, the *operation* of transport accounts for around two-thirds of total energy use (see Table 3.1). The remaining one-third is related to peripheral activities, for example the manufacture and maintenance of vehicles and infrastructure.

A similar picture emerges when one examines not energy use but greenhouse gas emissions. For example in the lifetime of a typical car, fuel consumed in operation accounts for 70 per cent of total greenhouse emissions, with the remaining contributions coming from fuel production, car manufacture and car disposal (6 per cent each), and CFC releases (12 per cent) (Martin and Michaelis, 1992).

It is important at this stage to distinguish between *delivered energy* and *primary energy*. The amount of delivered energy consumed by a transport system is the quantity of energy consumed at the point of use,

Table 3.1 Transport energy demand in Britain

Activity	Energy use (TJ)	% of total
Vehicle operation	1,563.1	66–72
Vehicle manufacture	40.0–157.0	2–7
Raw material production	108.1	5
Vehicle maintenance	94.2	4
Infrastructure provision	32.7	1
Energy generation	344.2–423.2	14–19
Total	**2,182.3–2,378.3**	**100**

Source Howard, 1990

which takes no account of the energy costs involved in providing the fuel. A more representative measure is primary energy, which takes into account the energy demand of obtaining, processing and delivering the fuel. (Primary energy does not, however, take the comprehensive view of transport as described above, in which all processes, including infrastructure provision, raw materials extraction and vehicle manufacture, are taken into account.)

The energy consumed by a vehicle at the point of use – the delivered energy – can be considerably less than the overall primary energy consumption. As Table 3.1 shows, primary energy consumption accounts for between 80 and 91 per cent of all energy demand in transport. The discrepancy between primary and delivered energy is particularly large in forms of transport which are powered by electricity, as much of the primary energy is wasted as a result of the electromechanical and thermodynamic inefficiencies inherent in the generation and transmission processes. In particular, the average efficiency with which steam turbine-type power stations produce electricity from their original fuel is between 30 and 35 per cent, with further losses incurred during transmission.[1]

In the case of vehicles powered by internal combustion (IC) engines, primary energy demand is comprised of the energy consumed at the point of use plus the energy required to refine and deliver the petroleum-based fuels. The major component of energy consumption is in the vehicle itself, rather than at the power station as in the case of electric vehicles.

Interestingly, the conversion of energy to traction is more efficient in electric vehicles than in IC-engined vehicles, but, as noted above, the benefit is lost as a result of the large proportion of energy wasted during the electricity generation process. Estimates have been made of the relative efficiencies of comparable electric and IC-engined vehicles. The efficiency with which an electric vehicle (EV) converts mains electricity to propulsive work is around 70 per cent, compared with just 15 per cent for a conventional IC-engined vehicle. When fuel production processes are taken into account, the overall propulsion efficiency of the EV drops to 19 per cent, whilst that of the IC-engined vehicle is reduced only marginally to 13 per cent (Francis and Woollacott, 1981).

In the era of steam railways, coal was the principal transport fuel in developed countries. During the 20th century, oil, in the form of petroleum, has gradually replaced coal. Transport now derives its energy almost exclusively from oil. In Britain, for example, over 99 per cent of transport's delivered energy is derived from petroleum.[2] Although the various forms of petroleum fuel have different energy demands associated with their production, these are minor when compared with the exceptionally high energy loss that occurs in electricity production.

Transport's share of CO_2 emissions

Global emissions of CO_2 are currently in the region of 27,000 million tonnes (Department of the Environment, 1993). In 1991, total anthropogenic emissions of CO_2 in the UK amounted to around 583 million tonnes, of which around 24 per cent was produced by the operation of passenger and freight transport (see Figure 3.1). A similar 'transport' percentage is found throughout the rest of Europe, as well as North America, Japan and Australia (OECD, 1991).[3] Figure 3.2 shows how the 'transport' segment of CO_2 emissions in Britain is itself divided between different types, or modes, of transport. Around two-thirds of CO_2 emissions are produced by passenger transport, as distinct from freight.

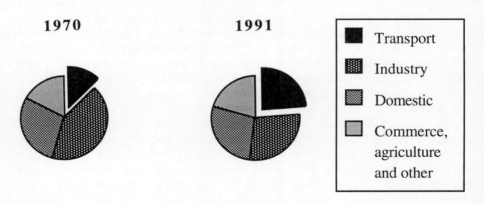

1970   **1991**

■ Transport

▨ Industry

▩ Domestic

□ Commerce, agriculture and other

Total: 667 million tonnes   **Total: 583 million tonnes**

Source Department of the Environment, 1993

Figure 3.1 CO_2 emissions from different end-uses in Britain

Within the spectrum of transport modes, two forms of travel can be identified as being distinctly separate in nature from the rest. These are the non-motorized modes of walking and cycling. Being powered entirely by human effort, they are fuelled by the complex system of energy production within the human body. Although human respiration is a producer of at least three greenhouse gases – carbon dioxide, water vapour and methane – it is assumed here that this source of emissions is negligible, since humans, together with the rest of the animal and plant kingdoms, form part of the natural carbon cycle (see Chapter 1).

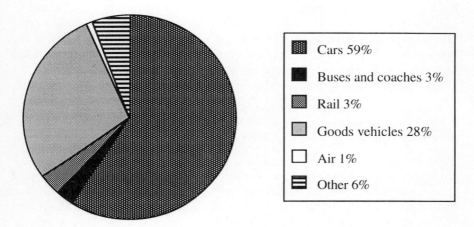

Cars 59%

Buses and coaches 3%

Rail 3%

Goods vehicles 28%

Air 1%

Other 6%

Note The chart refers to the *operation* of transport, and excludes peripheral activities such as those detailed in Table 3.1
Source Hughes, 1992

Figure 3.2 CO_2 emissions by different transport modes in Britain

This book examines the passenger subsector of transport, including all the significant modes of personal travel – car, rail, cycle, walking, and bus and coach. It includes domestic air travel, but not waterborne travel modes such as ferry, hovercraft and hydrofoil, which are of little overall significance for passenger travel and related CO_2 emissions.

'Buses and coaches' include public service vehicles of all sizes and for all operations, and cover the whole range of services from urban bus operations to long-distance coach travel. 'Rail' includes national railway networks together with urban rail operations, plus light rail transit (LRT) or tram systems of the type found in European cities like Zurich and Amsterdam. LRT typically operates on a mixture of streets, converted rail lines and newly built track. Tram networks have recently begun to return to cities in the USA and Britain. The first stage of a city-wide tram network opened in Manchester in 1992, and a similar scheme is under construction in Sheffield. Systems are also planned for Bristol, Edinburgh and Bimingham.

'Air' refers to domestic aircraft operations only, because this book is not concerned with travel between countries – only within them. In 1986, domestic journeys (those beginning and ending in the same country) accounted for just 4 per cent of the 90 billion passenger kilometres travelled from UK airports (Martin and Shock, 1989). Domestic flights therefore represent a small fraction of total air travel in Europe. In

North America, inter-city trips tend to be longer, for geographical reasons, and air travel holds a greater share of the market for these trips.

The growth of energy use in transport

Personal travel accounts for an increasing share of energy consumption and CO_2 emissions in developed countries. In 1970, passenger and freight transport operations accounted for 13 per cent of Britain's CO_2 emissions, whilst in 1991 their proportion had risen to 24 per cent, as Figure 3.1 illustrates. Energy consumption and CO_2 emissions in personal travel are determined by three variables:

- the *volume* of motorized travel;
- the *mode* of transport employed to undertake travel; and
- the *specific energy consumption* of the modes involved.

The net effect of these has been an historical increase in mobility, energy consumption and emissions. Since the Second World War, the developed world has witnessed a sustained growth in personal travel volume. Between 1952 and 1991, total travel in Britain grew from 223 to 689 billion passenger kilometres, more than a threefold increase (Department of Transport, 1992a). Moreover, Table 3.2 shows that Britain's historical growth in travel volume has involved an increase in both the *number* and the *length* of journeys undertaken. However, there is evidence that the number of journeys per person per week is stabilizing, with most of the growth in passenger mileage now attributable to a lengthening of trips. In the decade from 1975 to 1985 the number of journeys undertaken by the average Briton increased only marginally, from 18.2 to 18.5 per week, while the weekly travel distance increased by a third (Potter and Hughes, 1990).

Table 3.2 Journeys and travel distance per person in Britain, 1965 and 1985

	1965	1985
Journeys per person per week	11.2	13.2
Average journey length (km)	10.1	12.1
Travel distance per person per week (km)	112.6	160.9

Note the figures exclude journeys shorter than 1 mile
Source Department of Transport, 1988

The historical increase in Britain's personal travel volume has not been divided equally between different modes of transport. Figure 3.3 shows that the dominant effect has been a large growth in the use of cars – which, as it will be seen, are among the least energy-efficient forms of transport. The increase has been particularly rapid for leisure-related trips, which now account for around a third of all kilometres travelled. The rise in car use can be related to a growth in car ownership: between

1970 and 1990, the number of cars in Britain increased from 2.2 million to 19.7 million, and currently stands at around 376 per thousand people – a fairly typical European average (Department of Transport, 1993a). Meanwhile the volume of travel by rail changed very little, and the use of buses, coaches and pedal cycles declined. There has been a strong increase in air travel, but from a very small base.

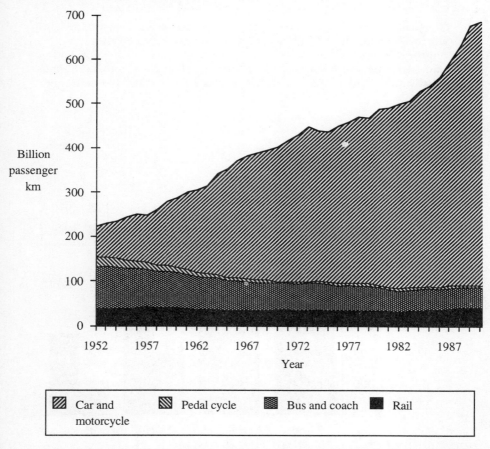

Note The graph excludes 'walk' trips. The 1985 National Travel Survey shows that a person walks on average 8 km per week: this leads to a total annual national figure of around 23 billion km.
Source Department of Transport, 1992a

Figure 3.3 The growth in personal travel in Britain

Throughout the developed world, a similar pattern of rising car ownership and personal mobility has taken place. Since 1970, the total number of cars in OECD countries has doubled, while the growth in car traffic, particularly in Europe, has followed closely behind (OECD, 1991). In Britain, the average person travels around 12,000 km per year.

Comparable travel intensities are found in other European countries including France, Germany, Italy and the Netherlands. In the USA, the average is higher, at 17,200 km, while Japan is well below European levels at 8,300 km per person (Department of Transport, 1992a). Figure 3.4 shows how car travel has increased throughout the OECD in the last 20 years. In all areas of the developed world, car use per capita has increased strongly. In the words of Martin and Michaelis (1992):

> The process of motorisation appears to be occurring everywhere and saturation in car ownership levels is beginning to be seen in the richest countries. In the US there is one car for over 600 in every 1,000 population – more than the number of drivers. If this ownership level is to be attained globally, the world is destined to have at least 10 times the current number of cars.

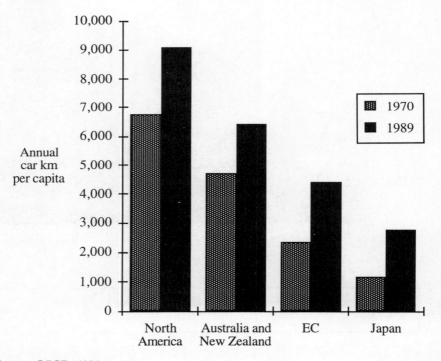

Source OECD, 1991

Figure 3.4 Car use per capita in selected regions, 1970–89

It is not only the number of kilometres travelled that is forcing up energy consumption and CO_2 emissions. The relative energy consumption of different travel modes – in terms of the amount of energy required to carry one person for one kilometre – also has a fundamental effect on overall energy use in transport. *Specific energy consumption* (SEC), measured in megajoules per passenger kilometre (MJ/pkm), is a function of two factors: firstly *vehicle fuel economy*, measured in megajoules per vehicle kilometre (MJ/vkm), and secondly *occupancy*, or the number of persons carried by a vehicle.

The average occupancy of vehicles is highly variable, given the enormous range of loads experienced by some modes. In particular, buses and trains may be operated almost empty at some times, whilst on other occasions the load factor may be well over 100 per cent as a result of additional standing passengers.[4] In the period 1970 to 1990, the average number of passengers carried by a car in Britain declined from 1.86 to 1.70, whilst on buses the average occupancy fell from 14.8 to 8.7. On railways, the average number of passengers carried by a train was unchanged at 98 (Department of Transport, 1992a).

Since the 1950s, specific energy consumption – the amount of energy consumed in carrying one passenger for one kilometre – has hardly changed at all for cars in Britain, despite major advances in vehicle technology. A decline in load factors, coupled with the absence of any substantial improvement in automobile fuel economy explains this lack of progress.

Meanwhile the railways have experienced an enormous improvement in SEC, which has fallen from over 6.5 MJ/pkm in 1950 to less than 2.0 MJ/pkm in 1990. Most of the change took place between 1950 and 1965, as modern diesel locomotives replaced relatively inefficient steam engines. More recent gains have been achieved as a result of technological improvements in the rail stock and the use of smaller trains on lightly used routes.

The SEC of bus travel has more than doubled since 1950, as a result of two factors. Firstly, in recent years there has been a trend towards smaller vehicles: although these typically have better fuel economy, they carry proportionately fewer passengers than a large vehicle. Secondly, the patronage of buses has declined, leading to lower average load factors.

By contrast, a striking reduction has taken place in the SEC of air travel, which improved by 75 per cent between 1963 and 1986 (although the trend has since levelled off). The gain is attributable not only to advances in aircraft technology, which are discussed in more detail later in the chapter, but also to an increase in occupancy, which has been the result of both larger aircraft and increased load factors (Martin and Shock, 1989).

There is a wide variation in SEC between different travel modes. Figure 3.5 illustrates the primary energy consumption of various modes currently in operation in Britain, under both 'typical' and '100 per cent' loading conditions. Aircraft and large cars rate as the most energy-intensive modes, while double-decker buses and LRT systems are among the most energy efficient. In the rail sector, diesel trains generally use less energy per passenger kilometre than electric, as a result of large inefficiencies that consume energy in the process of generating electricity. For all motorized travel modes, the difference between typical and maximum load factor, represented by the ratio of the dark area to the whole area, is substantial.

The aspects of energy demand described above are summarized schematically in Figure 3.6. It indicates how different characteristics combine to determine the overall energy consumption of a particular form of travel. This representation is a simplification of real life, in which there are other influences and feedbacks between the factors itemized in the diagram. In particular, it is widely known that the choice of travel mode has an influence on both the frequency and length of journeys, and therefore affects more than just the specific fuel economy.

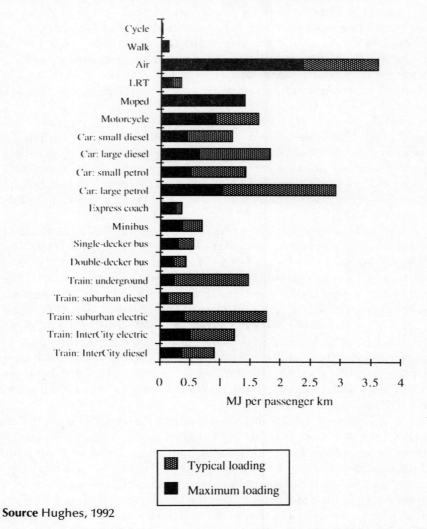

Source Hughes, 1992

Figure 3.5 Primary energy requirements of different travel modes in Britain

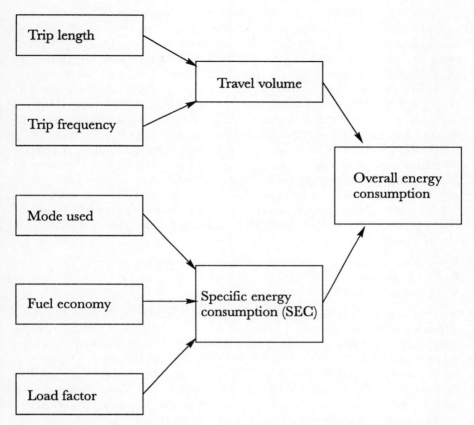

Figure 3.6 The influences acting upon transport energy demand

The influence of land-use patterns

The historical increase that has taken place in personal travel has been the result of several trends. Since the Second World War, car ownership has grown substantially, as discussed above. Alongside this trend, the real price of motor fuel has not increased significantly other than in the form of short-term fluctuations. Although these factors are important influences in themselves, they have also contributed to a lengthening of journeys via other, indirect means. In particular, a long-term trend towards dispersed settlement patterns has been a side-effect of these changes in travel behaviour.

Land-use patterns have a direct influence on all the areas represented in Figure 3.6. The spatial separation of homes, shops, workplaces and leisure centres determines journey lengths, and hence travel volume. Settlement patterns also influence the modes of transport that are used: in particular, dispersed land-use patterns tend to encourage further dependence upon the car for everyday travel, leading to a 'vicious circle' of motorization. Broadly speaking, the growth in ownership of cars, coupled with the sustained availability of cheap fuel on which to run them, has enabled people to adopt more mobility-intensive lifestyles, and obliged others to do the same. Homes have moved away from the central areas of towns and cities towards the periphery. Meanwhile leisure and retail facilities have taken advantage of cheap and freely available land on the outskirts of towns by opening large complexes that are difficult to reach other than by car. Similar trends have taken place with schools and hospitals. In the 1990s, many of those who have until now managed without a car are effectively being forced into acquiring one, in order to gain access to an increasingly dispersed collection of facilities which would at one time have been accessible on foot or by public transport.

Land-use planning thus exerts a powerful influence on the amount of energy consumed for travel purposes. Three characteristics in particular are of interest:

- Urban density;
- Settlement size; and
- Urban structure and centralization.

As one might expect, there is a definite correlation between urban density and travel demand. Higher-density settlements tend to have more opportunities for the use of local facilities, so the net demand for travel tends to be relatively low. In addition, density has a direct influence on the modes of travel employed by residents: space shortages tend to discourage widespread car ownership, while the large numbers of people using particular transport corridors create ideal conditions in which public transport services can flourish (Departments of the Environment and Transport, 1993).

There is a large volume of evidence supporting the idea that dense settlements tend to reduce travel demand and to encourage the use of public and non-motorized transport. Table 3.3 compares the distances travelled by residents of different population densities in Britain, and reveals a rapid decline in car use as density increases. A survey of cities world wide has revealed a strongly inverse correlation between urban density and gasoline consumption,[5] from Phoenix, with fewer than 10 persons per hectare (ppha), to Tokyo, with more than 100. North American and Australian cities typically lie in the range 10–20 ppha, while most European cities range from 30 to 70 (Newman and Kenworthy, 1989).

Potter and Hughes (1990) have used travel survey data to demonstrate a relationship between population density and walking distances to local facilities, as well as modal distribution. It is shown that there are critical density thresholds at 15–20 ppha and 45–50 ppha. Above the former threshold, walk trips rise to approximately 35 per cent of all trips, whilst above the latter walking rises to 40 per cent of trips. Similar thresholds can be identified for bus, rail and car use.

Table 3.3 The relationship between urban density
and travel behaviour

Density		Kilometres per person per week				
(persons per hectare)	All modes	Car	Local bus	Rail	Walk	Other
Under 1.00	206.3	159.3	5.2	8.9	4.0	28.8
1.00–4.99	190.5	146.7	7.7	9.1	4.9	21.9
5.00–14.99	176.2	131.7	8.6	12.3	5.3	18.2
15.00–29.99	152.6	105.4	9.6	10.2	6.6	20.6
30.00–49.99	143.2	100.4	9.9	10.8	6.4	15.5
50.00 and over	129.2	79.9	11.9	15.2	6.7	15.4

Source Department of Transport, 1988

Settlement size has a less obvious influence on travel behaviour.
Although large cities and conurbations tend to be associated with lower
travel intensities, there are awkward exceptions such as Greater London,
where personal travel demand is significantly greater than in similarly
sized settlements in Britain (Figure 3.7). It appears likely that variations
in income provide at least a partial explanation for this anomaly.

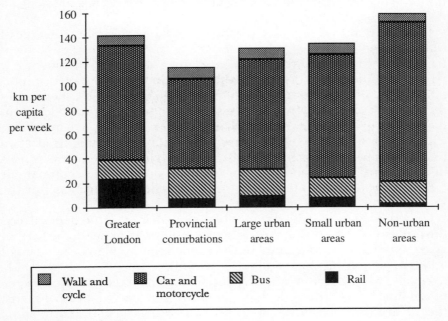

Source Maltby et al, 1978

Figure 3.7 Travel by residents of different settlement sizes, 1972

Any correlation between settlement size and travel demand may be
partly explained by the fact that larger settlements are likely to be more
self-contained in terms of access to people and services, and more able to

support public transport facilities. However, there is not a clear linear relationship between urban size and average trip length, because journey lengths also tend to increase with settlement size (Departments of the Environment and Transport, 1993).

Since the 1960s the population of Great Britain has been moving away from urban areas and into smaller settlements (Champion, 1987a) – a phenomenon known as 'counter-urbanization'. In 1985, 51 per cent of Britons lived in settlements of population 100,000 or more, while a further 38 per cent lived in settlements of between 3,000 and 100,000 people. The remaining 11 per cent lived in 'rural' settlements, of fewer than 3,000 people. Counter-urbanization has increased the number of people living in small, rural settlements which are not large enough to support the employment and social opportunities necessary for self-containment. In this way, the gradual exodus from large, traditional settlements has exacerbated the demand for travel, especially by private car.

The influence of settlement size upon energy consumption appears to act primarily through travel mode and journey length, rather than through journey frequency (Owens, 1986). The number of journeys undertaken per person shows very little change between settlements of different sizes (Maltby et al, 1978).

Finally, urban and regional structure – and in particular the degree of centralization – have a significant influence on travel demand. Many of the activities traditionally found in city centres have begun to move away from central locations, as geographical proximity becomes less important for business, and rising car ownership makes peripheral sites increasingly accessible. In some cases, demand for out-of-town locations has pushed ground rents on the urban periphery up to levels comparable with those found in the town centre. Manifestations of decentralization include business parks, and out-of-town shopping and leisure centres – developments which tend to require longer journeys than comparable facilities located in traditional town centre sites (Departments of the Environment and Transport, 1993).

On a regional scale, the distance between adjacent settlements affects the number of trips that are made between them. In general, the further apart settlements lie, the more self-contained they tend to be in terms of access to people, products and services (ibid).

Finally, while there is an unmistakable inverse relationship between urban density and travel demand, it would be a mistake to assume that the two are inevitably linked. Most towns and cities do fit the correlation, but it is nevertheless possible to design relatively low-density towns and cities that nevertheless have a healthy public transport network and a low level of car use. Hall (1991) holds the view that urban structure, rather than density, is the key determinant of energy consumption. For example, '... the efficiency of a city like Stockholm is due not to its relatively high overall density but to its polynuclear structure'.

Socio-economic influences on travel demand

The previous sections have documented the growth in personal travel that has taken place over the last few decades, and described some of the aspects of land-use planning that have contributed to this growth. Underlying these trends are a number of socio-economic factors, including the following:

- Economies of scale in the manufacturing sector have led to increased specialization of production. The result is an increasingly specialized workforce, which is prepared to travel further in order to find employment.
- Economies of agglomeration, particularly for manufacturing industry, are diminishing, with the result that centres of employment are becoming increasingly decentralized.
- The recent growth of the service sector has tended to increase the amount of travel undertaken in the course of business.
- Advances in communications technology have made it possible for different functions within businesses to be geographically separated, leading to a dispersal of office-based employment.
- Rising incomes have enabled more people to acquire cars, and allowed people to adopt more travel-intensive lifestyles.
- A growing number of women are in full-time employment, increasing the number of work-related trips and making it more difficult for families to live close to the workplace of both adults.
- Recent demographic changes have included an increase in the number of people living alone, who rely on travel for social interaction, and in the number of retired people, whose leisure activities are a major generator of travel.

(Departments of the Environment and Transport, 1993.)

It is clear, therefore, that the present trend towards increased mobility and away from centralized urban structures is the result of many factors acting together. The growth in mobility, decentralization, energy use and emissions is in many ways self-sustaining. As rising incomes allow more people to acquire cars, it becomes possible for businesses and services to move to more peripheral locations, which in turn require people to use cars to reach them. The widespread acquisition of cars allows residential areas to develop in remote locations, served by an expanded network of motorways and trunk roads. The better access provided by these new roads encourages car users to travel further and more frequently. Meanwhile the dispersed pattern of land use that emerges is hostile to public transport, which in many cases ceases to become commercially viable.

The wider view: non-operational energy demand in personal travel

As explained earlier, personal travel contributes to CO_2 emissions not only through the combustion of fuels used in propelling the vehicles, but also as a result of a variety of peripheral activities. Principal among these are the manufacture and maintenance of vehicles and the infrastructure on which they operate. These 'secondary' sources account for up to a sixth of all energy used in passenger transport, as detailed earlier in Table 3.1.

Within Britain's manufacturing sector, the vehicle and transport equipment (VTE) subsector is estimated to be the third largest consumer of energy, with an estimated consumption in 1990 of between 122 and 128 petajoules (PJ).[6] Energy consumption within the VTE subsector is not dominated by any single fuel, but is spread between coal, oil, natural gas and electricity. Surprisingly, over half the energy consumed by the VTE industry is used for the space-heating of commercial premises (Department of Energy, 1984). But compared with other manufacturing industries, VTE is not especially energy intensive. In 1979 it had an energy intensity of 18.7 megajoules per pound sterling of net output, compared with 25.9 MJ/£ for the food and drink industry and 237 MJ/£ for iron and steel manufacturing.

One particular part of the manufacturing sector, the car industry, is heavily dependent upon more energy-intensive industries, particularly iron and steel processing. When one considers not only the vehicle manufacturing industry but all the heavy industries that supply it with components, it becomes clear that as much as two-thirds of a car's energy content is accounted for by the production of the iron and steel used in its construction (see Figure 3.8).

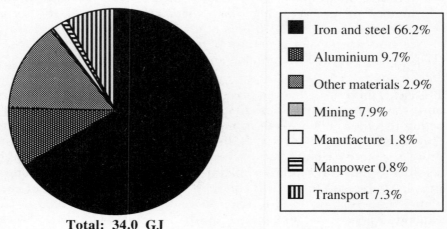

Total: 34.0 GJ

Iron and steel 66.2%

Aluminium 9.7%

Other materials 2.9%

Mining 7.9%

Manufacture 1.8%

Manpower 0.8%

Transport 7.3%

Source SMMT, 1980

Figure 3.8 Energy consumed in the production of a car

Another significant demand for energy relates to the construction and upkeep of transport infrastructure such as roads and railway tracks. Zeevenhooven (1990) has estimated the energy costs of building and maintaining road and rail infrastructure in Sweden. He shows that transport by rail is less energy intensive, in terms of infrastructure and maintenance, than road transport. Meanwhile, perhaps not surprisingly, air travel infrastructure requires relatively very small amounts of energy per passenger kilometre.

In summary, although most energy consumed by transport is used in the operation of vehicles, the energy demand of peripheral activities, such as the construction and maintenance of vehicles and infrastructure, should not be overlooked. These categories of energy demand could be particularly important when, for example, assessing options for a new transport system, where the choice of mode has implications not only for the consumption of energy in operation, but for the construction of new infrastructure.

Energy use and greenhouse emissions

Recent trends in personal travel have been characterized by a sustained increase in passenger mileage by the least energy efficient of travel modes, the private car and the aircraft (although air transport still represents a very small percentage of domestic travel). This upward trend has been translated into a corresponding increase in fossil fuel consumption and emissions of CO_2.

As noted in Chapter 2, CO_2 is the most significant of anthropogenic greenhouse gases. For any fossil fuel, emissions of this gas are directly proportional to fuel consumption. Historically the trend towards petroleum fuels away from coal has to some extent offset the growth in CO_2 emissions, because oil has a lower carbon content than coal, and also because internal combustion engines are substantially more energy efficient than their predecessor, the steam engine. But the sheer growth in transport demand – a 'mobility revolution' – has greatly outweighed the improvement in the efficiency of individual vehicles. CO_2 emissions are rising as the relentless growth in travel demand continues.

A number of other transport-related emissions, besides CO_2, are believed to contribute, directly or indirectly, to global warming. In particular, leakage of CFCs from vehicle air conditioning systems is a direct source of greenhouse emissions. This source of CFCs is particularly important in North America, where air conditioning is a feature of most production cars. By contrast, it is less widespread in Europe. Other applications of CFCs include foam polymers used in vehicle trim, coolants for electrical transformers, and solvents used to clean electronic components. The Montreal Protocol (see Chapter 2) has secured an agreement for CFCs to be eliminated altogether from most applications, and it seems certain that substitute gases will be found for all these applications.

The role of carbon monoxide (CO) should not be overlooked. Although this gas has no direct radiative forcing effect, its reaction with hydroxyl (OH) radicals has the effect of allowing the abundance of other greenhouse gases, such as methane, to increase (Ramanathan, 1988).[7] Vehicles can also contribute directly to emissions of methane, through the release of hydrocarbons (HC) from fuel tanks. Leakage from vehicles running on natural gas also constitutes a potential source of methane emissions, since methane is the main constituent of this fuel. Motor vehicles have been estimated to contribute around 1 per cent of overall methane emissions (Department of Energy, 1990).

At present, methane emissions are the result primarily of agriculture, mining and landfill activities. But a proliferation of natural gas as a transport fuel would be likely to increase the overall output. It is conceivable that cars running on natural gas would produce more 'CO_2-equivalent' than conventional petroleum-driven vehicles, because of methane leakages occurring during the production and distribution of fuel (Mills et al, 1991).

Mention should also be made of the contribution of nitrogen oxide (NOx) emissions to global warming. One of the oxides of nitrogen, nitrous oxide (N_2O), is a powerful greenhouse gas, while the others contribute indirectly to global warming when they react in sunlight with hydrocarbons to form ground-level ozone. Although N_2O is estimated to be responsible for only 5 per cent of greenhouse warming, transport is the largest producer of this gas in Britain.[8] Table 3.4 gives details of transport's emissions of CO, NOx and HC in the UK.

Table 3.4 Transport's emission of regulated pollutants in the UK

Pollutant	Emissions from transport (thousand tonnes)	% of UK total
Nitrogen oxides (NOx)	1,646	59
Carbon monoxide (CO)	6,060	90
Hydrocarbons (HC) (excluding methane)	1,300	49

Note These figures refer to both passenger and freight transport, as separate data are not available
Source Department of the Environment, 1992

Regulated pollutants and EC Directives

Emissions of 'regulated' pollutants (NOx, CO and HC) from road vehicles in Europe, as in North America, are strictly controlled by legislation. Before June 1989, it was expected that small and medium cars sold in the European Community would be able to meet the forthcoming emissions requirements using a combination of lean-burn combustion and simple oxidation catalysts. But the legislation for small cars that was introduced in June 1989 was more stringent than most experts in the industry expected, effectively requiring all new cars sold after the end of 1992 to be fitted with three-way catalytic converters (Boehmer-Christiansen, 1990). Three-way catalysts contain a ceramic honeycomb impregnated with catalytic metals, principally platinum, which oxidize carbon monoxide and unburnt hydrocarbons while reducing nitrogen oxides. The main products are nitrogen, water vapour and carbon dioxide. Critically, the use of catalysts requires that engines be tuned to run at the 'stoichiometric' ratio of air to fuel, namely the ratio whereby fuel and oxygen are present in the precise quantities for complete combustion to take place, typically 14.7:1. By contrast, lean-burn engines achieve reduced emissions by operating on a much weaker mixture, with air to fuel ratios as high as 25:1. In effect, catalytic converters and lean-burn engines are mutually exclusive and the EC legislation, in the form of Directive 91/441, was designed to promote the use of catalysts.

EC Directive 91/441 marks the end of much uncertainty on the part of motor manufacturers on what the EC's favoured strategy for reducing emissions would be. The Commission has not dictated the type of technology – catalysts or lean-burn engines – that may be used by car manufacturers to achieve the prescribed standards for regulated pollutants. But the universal move towards catalytic converters and engines running at stoichiometric air–fuel ratios has dealt a near-fatal blow to the prospects of a European market for lean-burn cars.

In terms of CO_2 emissions, the Directive was misguided. By its very nature lean burn implies a more efficient use of fuel. Instead of forcing the exhaust gases through a catalyst-coated honeycomb, as in the case of a 'cat' car, lean burn uses a weakened fuel–air mixture to ensure more complete combustion, and less harmful exhaust gases. For vehicles of equivalent performance, emissions of CO_2 are some 10 per cent less from a lean-burn engine than from a catalyzed engine.

At the time of the 1989 legislation, lean burn was still at an early stage of development as a means of controlling emissions. Critics have argued that, given a few more years, lean burn technology would have been sufficiently advanced to meet the stringent pollution regulations set down by the EC, producing *at the same time* a reduction in fuel consumption and CO_2 emissions. It appears likely that lean-burn engines being developed in Japan will soon be capable of meeting the emissions requirements prescribed by the EC. Effectively, it may soon be possible to buy a new car that is not fitted with a catalyst, but which nevertheless satisfies current emission standards.

The EC has proposed more stringent emission limits to be introduced in 1996, corresponding approximately to legislation being introduced in the USA from 1994. Holman et al (1993) believe that the motor industry will have little difficulty in achieving the new EC limits using existing catalytic converter technology. It remains to be seen whether lean burn will also be capable of meeting the 1996 emissions standards.

In terms of greenhouse gas emissions, the reduction in output of the regulated pollutants resulting from the 1989 legislation will have a generally beneficial effect. However, a number of factors will act against this improvement.

Firstly, whilst individual vehicles will become less polluting, overall emissions of regulated pollutants may still increase as a result of rising traffic volumes. A sustained increase in vehicle mileage, such as that anticipated in Britain's National Road Traffic Forecasts (Department of Transport, 1989a), may wipe out altogether any benefit delivered by the introduction of catalytic converters (Holman et al, 1993).

Secondly, it is known that catalytic converters do not operate at full efficiency in cold conditions. Curiously, the EC test cycle that is used to measure exhaust emissions requires the test vehicle to be warmed to 25°C, in order to ensure that the catalyst is working properly. But an investigation by Broom (1991) has shown that when the EC test cycle is performed at typical European air temperatures – which are significantly lower than 25°C – catalyst-equipped cars produce noticeably greater levels of all three regulated pollutants than the official test results suggest. Output of CO may even exceed the EC limit value as the low temperature renders the catalytic converter inoperative. The fact that catalysts commonly do not work for the first few kilometres of a journey means that short car journeys are significantly more polluting, per kilometre, than longer ones. In addition, cars fitted with catalysts are more susceptible to 'cold-start' emissions than those without. If travel patterns continue to move towards an increased proportion of short journeys, it is likely that the emission of regulated pollutants will increase accordingly. In an examination of the cold-start penalty on emissions, Holman et al (1993) have estimated that by the year 2000, 50 per cent of carbon monoxide and hydrocarbon emissions from cars could be the result of cold starts.[9]

Thirdly, direct emissions of N_2O, a powerful greenhouse gas, have been shown to be between five and eight times greater from cars fitted with catalytic converters than from cars without. But the picture is muddied somewhat by the fact that 'indirect' emissions of N_2O, derived from nitrogen monoxide (NO) and nitrogen dioxide (NO_2) in vehicle exhausts, may be less in the case of catalyst-equipped cars. The overall effect is unclear, and further research is needed to understand the process in which N_2O is formed within catalytic converters (Gould and Gribbin, 1989).

Finally, three-way catalytic converters are known to have a detrimental effect on vehicle fuel economy. For example, the official urban fuel consumption figures for a 1.4 litre Ford Escort in 1990 were 8.9 litres per 100 km for the uncatalyzed model and 10.0 litres per 100 km for the catalyst-equipped version – a shortfall of 12 per cent. Similarly, tests undertaken by Rover Group on two identical Metro cars – one with a catalyst and one running lean burn – revealed a fuel economy difference of 8 per cent (Broom, 1991). If vehicles are using more fuel per kilometre as a result of being fitted with a catalyst, there will be a resultant increase in CO_2 emissions. Further energy costs are associated with the manufacture of catalytic converters.

In summary, the progressive introduction of catalytic converters to the world's car fleet will, on an individual vehicle basis, reduce the emissions of certain greenhouse gases, as well as gases that contribute indirectly to global warming. But emissions of CO_2 will increase significantly – though the effect could be counteracted by improvements in other areas of vehicle technology. The production of 'regulated' pollutants will not be entirely eliminated by catalysts, and is likely to rise if traffic growth continues.

This book does not attempt to predict the effects of different policies on all greenhouse gases; instead, the focus will be on CO_2 alone. With a few exceptions, reductions in emissions of CO_2 are generally accompanied by reduced emissions of other greenhouse gases – usually because less fossil fuel is being consumed. In many cases CO_2 can therefore be regarded as a 'surrogate' for these other gases.

The influence of fuel consumption on greenhouse emissions

Advances in vehicle technology have made possible substantial improvements in vehicle fuel economy. But these improvements have often been accompanied by declining load factors, with the two trends to some extent counteracting one another. For this reason, it is useful to measure energy consumption per person carried, in terms of SEC. As explained earlier in this chapter, this measure is a function of two variables: firstly the occupancy of the vehicle; and secondly its fuel consumption, measured in MJ/vkm. In the airline industry, for example, there has been a trend towards larger and more efficient aircraft. Whilst these consume more fuel per kilometre than their predecessors, the energy consumption per passenger kilometre is less.

The growth in energy demand arising from an increase in travel volume, as described earlier, can be offset by improvements in SEC. The following sections will focus on the factors that determine fuel consumption in the motorized travel modes – cars, buses, coaches, trains and aircraft.

The energy efficiency of the non-motorized travel modes, walking and cycling, can to some extent be improved through the development of more efficient machinery and infrastructure. But the benefits of such advances are not considered to have a direct effect on greenhouse emissions. Rather their value lies in the improvement in travel conditions that they bring for the user, encouraging modal transfers from less energy-efficient forms of transport.

Fuel economy in cars

In order to understand the nature of energy use in cars, it is helpful to divide a vehicle's energy consumption into different 'sinks'. Figure 3.9 shows that typically less than a fifth of a car's energy consumption is used in providing motion. The remaining 82 per cent is lost – either as waste heat generated by frictional forces between moving parts, or as a result of the thermodynamic constraints associated with heat engines.

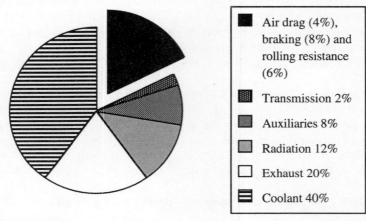

- ■ Air drag (4%), braking (8%) and rolling resistance (6%)
- ▦ Transmission 2%
- ▨ Auxiliaries 8%
- ▥ Radiation 12%
- ☐ Exhaust 20%
- ☰ Coolant 40%

Source OECD, 1982

Figure 3.9 Energy losses in a typical car journey

The resources invested by motor manufacturers in the development of vehicle technology are immense, and the results of this research have led to considerable advances in areas such as engine and transmission efficiency, materials technology and aerodynamics. In simple terms, improvements in vehicle efficiency can be used in two ways. On the one hand they can be used to improve the fuel economy, or 'miles per gallon', of the vehicle, by reducing its engine size, whilst leaving its performance unchanged. On the other, they can be used to increase the performance of the vehicle whilst leaving its engine capacity and fuel economy unchanged.

With the exception of the periods of short-lived frugality that followed the sudden oil price rises of the 1970s and 1980s, the second of these has been the dominant effect in most countries. In the presence of sustained low fuel prices, energy-efficiency technologies have generally been used to produce more powerful cars, whose fuel economy is virtually unchanged. An indication of this effect can be found in car advertising material, which commonly emphasizes performance in preference to fuel economy.[10]

But it would be wrong to suppose that motor manufacturers are uninterested in fuel economy as a potential marketing tool. It is reported that immediately after the Allied intervention in Kuwait in January 1991, Ford rapidly shifted the focus of its billboard advertising campaign in Britain away from large, powerful cars to the smaller, more economical models in its range, fearing that an imminent oil price shock would decimate the market for uneconomical cars. Similarly, Britain's 1993 Budget, in which a 10 per cent increase in motor fuel taxes was announced, resulted in a rash of car advertisements appearing in the national press, emphasizing miles per gallon rather than miles per hour.

Another important influence on car fuel consumption, in Britain at least, has traditionally been the 'company car factor'. The provision of company cars for the use of employees is a popular practice among British firms, and Britain has the highest proportion – around 70 per cent – of company-financed cars in the world. A car is regarded as a form of income-in-kind on which a reduced rate of tax is payable, and as such provides an economic benefit to companies (Potter, 1991). The company car factor has important implications for emissions of CO_2, both because company cars tend to have larger engine capacities than privately purchased cars, and because they tend to be driven further as a result of the fuel subsidy.

However, the scale of the subsidy given to company motorists has recently been declining, under pressure from environmentalists and economists alike, and it is probable that the 'incentive to drive' offered by discounted cars, mileage thresholds and free fuel will be largely eliminated. In his 1993 Budget speech, the Chancellor of the Exchequer announced that the 'scale charges' – the notional value of a company car for tax purposes, which has traditionally been underestimated by a considerable margin – would be raised to reflect the car's true value. Tax would then be payable on the car's list price, rather than in relation to engine size as before.

Following this substantial review of the company car subsidy, the influential Lex survey of company motoring revealed that as many as one in five company motorists in Britain was considering giving up the company car as a result of the new regime. Moreover, most of the drivers surveyed believed that the new arrangements were fairer than the old ones (Lex, 1993).

In Europe and North America, car manufacturers are required to submit all new models to a series of tests to establish their average fuel economy. In Europe the fuel economy data comprise three figures, corresponding to (i) an urban cycle, (ii) constant 90 kmh driving, and (iii) constant 120 kmh driving. The figures are intended as a guide for purchasers who wish to compare the fuel economy of different models. But as an indicator of real-life fuel economy they are less reliable. To illustrate this point, it is possible to compare the 'official' and 'actual' fuel economy of the car fleet as whole, and examine the gap between them.

The 'official' average fuel economy of the British car fleet has been calculated each year since 1978 by the Department of Transport using a sales-weighted average of the official data. This is given in Figure 3.10, which also shows average fuel economy calculated using an 'empirical' method, whereby the total consumption of motor spirit for each year is simply divided into the total annual car mileage. An average discrepancy of 22 per cent exists between the fuel economy derived from the official test results and that based on the aggregate volume of fuel actually used. Furthermore, the difference increased from 18 per cent in 1978 to 30 per cent in 1988. This 'mpg gap' is attributable to the shortcomings of the official tests as a representation of real-life driving conditions. The Department acknowledges this discrepancy, noting: 'The standard tests...cannot be fully representative of real-life driving conditions...The fuel consumption achieved on the road will not necessarily accord with the official test results' (Department of Transport, 1990).

The 'mpg gap' is important in terms of transport policy. According to the official measure, fuel economy improved by 17 per cent in the period 1978–87. But the *actual* improvement, as revealed by aggregate data, was much smaller – only 9 per cent.

A similar discrepancy of around 15 per cent has been observed in the US car fleet, where it has been estimated that actual fuel economy could fall to 30 per cent below the official ratings by the year 2010, principally as a result of growing urban congestion, greater highway speeds, and a rising proportion of car travel being undertaken in urban areas (Westbrook and Patterson, 1989).

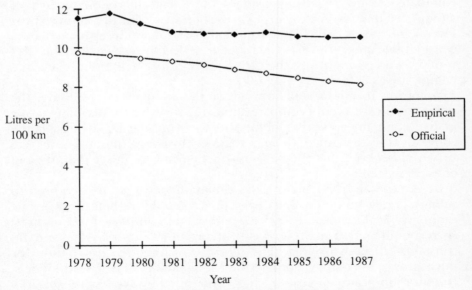

Sources National Audit Office, 1989; Department of Transport, 1992a

Figure 3.10 'Official' and 'empirical' estimates of average car fuel consumption

Factors influencing fuel consumption may be subdivided into *vehicle effects* and *operational effects*. Vehicle effects may be defined as the aspects of vehicle design that influence its fuel economy rating, including engine size, aerodynamics and body weight. Operational effects are the details of its usage, such as the type of roads on which it is normally used, the manner in which it is driven and the care with which it is maintained.

Vehicle effects

Almost all car engines fall into one of two categories: petrol (spark ignition) and diesel (compression ignition). Diesels are less powerful for a given engine capacity than petrol engines: the power rating for a diesel engine is between 20 and 40 kilowatts per litre, compared with 45 to 65 kW/litre for a petrol engine (Martin and Shock, 1989). However, diesel engines are more economical than petrol engines when cars of equivalent performance are compared. For example, a diesel-engined Vauxhall Cavalier sold in Britain has been shown to be between 4 and 22 per cent more economical than its petrol-engined counterpart (Redsell et al, 1988). Table 3.5 illustrates the fuel economy improvement associated with diesels across the whole car market, and shows that the benefit is particularly high in smaller cars. The world fuel economy record for a production car is currently held by the Citroen AX diesel, which recently achieved 2.5 litres per 100 km, beating the previous record set by a Daihatsu Charade (Cragg, 1992).[11] It should be noted that this record applies to production cars only; much higher levels of fuel economy have been achieved by purpose-built prototypes. For example, the winner of the 1991 Shell Mileage Marathon was a lightweight, highly aerodynamic machine that achieved a record 6142 miles per gallon or 0.05 l/100 km (Watkins, 1991).

Table 3.5 Average fuel consumption of petrol and diesel cars

Range	Fuel consumption (litres per 100 km)		% difference
	Petrol	Diesel	
Bottom	7.4	5.6	−25
Middle	8.8	6.8	−23
Top	10.5	8.4	−20
Average	**8.8**	**6.9**	**−23**

Source Lucas, 1990

The worse fuel economy of petrol engines compared with diesels can be related to three characteristics. Firstly, engine efficiency under partly-loaded conditions is poor in petrol engines, because the power output is controlled by throttling the fuel/air intake. Secondly, the ratio of fuel to air in petrol engines is held constant at approximately the stoichiometric value. In many circumstances better fuel economy could be achieved using a weaker fuel mixture. Finally, the compression ratio of petrol engines, which should be maximized for best fuel efficiency, is limited by the onset of spontaneous ignition ('knocking'). To raise the compression ratio requires an increase in the octane number of the fuel, which is influenced by refinery processing and the presence of anti-knock additives containing lead compounds. Legislation requires that all new cars in Europe and North America run on unleaded fuel, so the second route is not feasible (Martin and Shock, 1989).

In diesel engines the fuel is ignited not by a spark but by the compression of fuel within the combustion chamber. The superior fuel economy of diesel engines is the result of increased compression ratio and weaker fuel–air mixture. In addition, power output is controlled using fuel supply metering, rather than the less efficient technique of throttling with a fixed fuel–air ratio.

On the negative side, diesel engines create more noise and vibration than petrol engines, as a result of the greater compression ratio. They are also heavier and more expensive. Traditionally those fitted to cars have been of the *indirect injection* type. However, an increasing number of manufacturers are switching to *direct injection* because of its enhanced efficiency, although noise and emissions tend to be worse.

A number of technologies have been developed in the last two decades for improving the energy efficiency of petrol and diesel engines. As explained earlier, an increase in efficiency may be used either to improve fuel economy, or else to increase performance whilst maintaining the same fuel economy. Many of these technologies have been developed and marketed with increased performance in mind, rather than superior fuel economy. For example, *turbocharging* is a technique whereby exhaust gases are used to drive a fan which physically impels the fuel–air mixture into the cylinder. The result is a denser mixture, and a greater power to weight ratio. A turbocharger may be used in conjunction with *aftercooling*, whereby the air from the turbocharger is cooled using a heat exchanger so that more air is drawn into the engine at the induction stroke. *Intercooling* is the term given to the cooling of the fuel–air mixture before it enters the cylinder. As with turbocharging, this has the effect of increasing the density of the mixture, which has a beneficial effect on engine efficiency (Martin and Shock, 1989).

A study has been carried out by the Transport Research Laboratory comparing the fuel consumption of diesel and petrol cars under cold-start conditions (Pearce and Waters, 1980). The loss of fuel efficiency associated with cold running is shown to be less significant in a diesel engine than in a petrol engine. Moreover, the diesel engine takes less time to reach its full operating temperature. In short, the 'mpg gap' between actual and official fuel consumption, described earlier, is less significant for diesels than for petrol engines.

Fuel consumption is also strongly influenced by the *capacity* of the engine, or the total volume swept by the pistons inside the cylinders. In general, the bigger the engine, the higher the fuel consumption. Table 3.6 lists the average fuel economies of six engine size categories. A car of capacity 1.0 litre or less uses a little over half the fuel of a 2.0 litre car to cover the same distance.

Table 3.6 Average fuel economy of different engine sizes, 1986

Vehicle category		Average fuel economy (litres/100 km)
Petrol-engined		
Over	Not over	
	1000 cc	7.6
1000 cc	1200 cc	8.3
1200 cc	1500 cc	8.9
1500 cc	1800 cc	9.8
1800 cc	2000 cc	10.6
2000 cc		13.7
Diesel-engined		
Over	Not over	
1800 cc	2000cc	6.4
2000 cc		7.7

Source Martin and Shock, 1989

The *weight* of a vehicle, determined by the materials used in its construction, represents another major influence on fuel consumption. The principal components of a car's mass are the body (typically 28 per cent), the engine and transmission (21 per cent) and the trim, including glass (16 per cent). It is estimated that cars built from aluminium are around 30 per cent lighter than conventional steel-built vehicles. The fuel savings associated with this kind of weight reduction are estimated to be around 5 per cent for every 10 per cent cut in weight (Martin and Michaelis, 1992).

Table 3.7 lists the materials by weight that are found in a typical small car. Future advances in materials technology are likely to reduce the overall mass of iron and steel used in cars, in favour of non-ferrous metals and plastics. It is likely that aluminium will play an increasing role in vehicle manufacture, and could represent as much as 20 per cent of a car's weight by the year 2000, while plastics could reach 15 per cent (ibid). Pressure to produce lighter cars is particularly strong in the US market, which has strict fuel economy and emissions standards (Gooding, 1991).

The trend towards lighter cars is liable to be offset, however, by an increase in the amount of ancillary equipment, and in particular safety features. Tests conducted at the Transport Research Laboratory (1987) suggest that vehicle weight could be increased by around 3 per cent as a result of crash-resistant design features such as metal bars built into the passenger doors.

The weight of people and luggage carried by a car also influences its fuel consumption. This is not only because of the greater force needed to

Table 3.7 Materials used in the Austin Metro

Material	Mass (kg)	% of total
Steel	422.4	59
Cast iron	110.0	15
Aluminium alloys	20.5	3
Copper and its alloys	4.5	1
Zinc alloys	1.5	–
Lead alloys	9.2	1
Other metals	1.0	–
Rubber	35.0	5
Plastics	35.0	5
Glass	30.0	4
Other non-metallic	41.0	6
Total	**710.1**	**100.0**

Source Alexander and Street, 1989

accelerate the vehicle, but also because unless tyre pressures are readjusted, the rolling resistance of the vehicle increases when extra weight is added.[12] Martin and Shock (1989) estimate that each passenger of average weight is responsible for a 3 per cent increase in fuel consumption on main roads, and 4 per cent in urban driving conditions.

The transmission system – the assembly that carries power from the engine to the wheels through some form of gearing – is an important area of energy loss in cars. Fuel economy can be optimized by ensuring the best match between engine speed and driving conditions. Conversely, poor use of gears can lead to a fuel consumption penalty of up to 20 per cent (Martin and Michaelis, 1992). In recent years many more vehicles have been fitted with five-speed gearboxes rather than four-speed: the extra gear allows the engine to run more slowly, and therefore more economically, in high-speed driving.

Beyond conventional transmissions (both manual and automatic) lies the concept of *continuously variable transmission* (CVT), in which the conventional set of discrete gear ratios is replaced by an infinitely variable system. In this way the engine speed can be maintained at its most efficient level regardless of vehicle speed. CVTs were fitted to Daf cars in the 1970s, and have recently been reintroduced in small cars produced by other manufacturers, including Fiat, Ford, Rover and Subaru. The fuel saving resulting from the use of CVT is estimated to be as much as 20 per cent compared with a conventional four-speed gearbox (ibid).[13]

Meanwhile more advanced systems such as *automated manual transmission* have been developed, in which discrete gear changes are carried out electronically via a clutch rather than the fluid coupling device found in most automatic gearboxes. Fuel savings of up to 20 per cent have been claimed for these devices, compared with conventional four-speed transmissions (ibid).

As well as the mass of the vehicle and its contents, *aerodynamic drag* represents a significant loss of energy in cars, particularly in high-speed driving. It is dependent on the speed, frontal area and drag coefficient of the vehicle, according to the equation

$$D = 0.5\ C_d \rho\ S\ v^2$$

where D is the drag force, C_d is the coefficient of drag, ρ is the air density, S is the vehicle frontal area and v is the speed of the car relative to the air through which it is being driven. Vehicles currently in use have a typical drag coefficient of around 0.38, while highly streamlined prototypes have achieved values as low as 0.12 (ibid). Figure 3.11 breaks down the drag coefficient of a typical car into components, showing how various aspects of car design contribute to the overall drag coefficient Cd.

As discussed earlier, the effect of catalytic converters on fuel consumption is generally detrimental. By contrast, the *lean-burn approach* to emission control generally entails some improvement in fuel economy. Lean-burn engines, when fitted with no additional emission control equipment, consume around 10 per cent less fuel than a conventional engine.

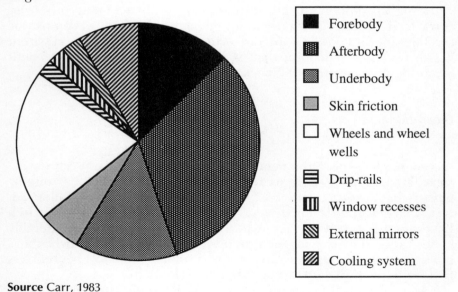

- Forebody
- Afterbody
- Underbody
- Skin friction
- Wheels and wheel wells
- Drip-rails
- Window recesses
- External mirrors
- Cooling system

Source Carr, 1983

Figure 3.11 Contributions to overall drag coefficient in a typical car

Electronic *engine management systems* are becoming increasingly common on production cars as part of the exhaust-cleaning operation. The most advanced of these receive data from various sensors in the engine, and use them to control the running of the engine. The system monitors such aspects as ignition timing, fuel injection, exhaust emissions, and even suspension and transmission control. Potentially, electronic management offers considerable scope for fuel saving by improving the efficiency with which a car is operated.

The age of a vehicle is also important when considering fuel economy. As a car gets older, worn components can begin to raise fuel consumption. Sources of inefficiency in the engine include ignition timing, carburation, slack belts and soiled filters. Tests carried out by the Royal Automobile Club in Britain, using pollution monitoring techniques developed by Donald Stedman at the University of Denver, Colorado, revealed that around 40 per cent of air pollution was produced by just 1 per cent of cars. Significantly, the vehicles that were in the worst condition were not necessarily the oldest (Cragg, 1992).[14] Other contributors to poor fuel economy include misaligned wheels, underinflated tyres, poorly adjusted brakes and poor-quality engine oils. The mandatory checks to which vehicles are subjected each year can, however, detect many of these faults.

The fuel consumption of a car can also be affected by the fitting of auxiliary equipment by the owner. Some accessories may affect the aerodynamics, as with roof racks, mirrors and body styling. A roof rack may increases the fuel consumption of a vehicle by up to 40 per cent at 120 kmh. In addition, the fitting of ancillary equipment may impose an extra load on the car's power output, as in the case of lamps, air conditioning and power-assisted steering. Typically the electrical system has an efficiency of no more than 10 per cent, so the net energy demand is considerable (Martin and Shock, 1989).

Operational effects

As one might expect, the manner in which a car is driven can have a considerable influence on its fuel consumption. Of particular importance is the rate of acceleration and deceleration, as well as overall speed. A comparison of different drivers carried out by the Transport Research Laboratory revealed that an 'expert' driver could return considerably better fuel economy than the group average. In urban, suburban and motorway tests, the 'expert' was able to consume 9, 10 and 24 per cent less fuel respectively than the rest of the group (Redsell et al, 1988).

Figure 3.12 illustrates the effect of speed on average fuel economy. At low speeds, poor fuel economy results from the frequent stops and starts associated with congested traffic conditions.[15] At the other end of the scale, fuel economy deteriorates as a result of higher speed and the greater wind resistance, which increases in proportion to the square of the speed.

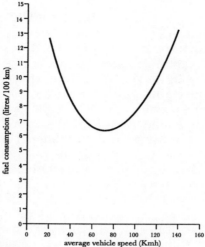

Source Redsell et al, 1988

Figure 3.12 The effect of average speed on fuel economy

The type of driving to which a car is subjected depends not only on the behaviour of the driver but also on the nature of the journey being undertaken. In particular, short car journeys are prone to poor fuel economy because a large proportion of the trip is covered with a cold engine. It has been shown that an average car typically takes 11 km to 'warm up' fully. During the first few minutes of warm-up, a car engine may use two to three times more fuel per kilometre than it would when fully heated (Pearce and Waters, 1980). In Britain, more than 60 per cent of car journeys are less than 8 km in length, of which many are made 'from cold' (Hughes, 1992). It has been estimated that the extra fuel consumed as a result of cold running accounts for as much as one-third of the total fuel consumed by cars in Britain (Armstrong, 1983).

As noted earlier, diesel-engined cars are less susceptible to cold-starting effects than cars with petrol engines. They are also more efficient in the low engine load conditions associated with short, urban trips.

A variety of technologies exist for 'preheating' car engines in order to reduce the fuel economy penalty associated with cold-starts. As well as electrically heated catalysts, a technique has been developed by Volkswagen where the engine's heat is stored in a highly insulated container filled with a fluid of high energy density, and then released again to warm the engine next time the car is used (Barrie, 1990). A similar system developed by Saab stores energy in crystals that change from solid to liquid when heated to 78°C, which are stored in a vacuum-insulated metal cylinder 33 cm long. The manufacturers claim that the device is capable of keeping the crystals above 78°C for a remarkable three days after the engine has been switched off (Griffiths, 1991).

Traffic management systems have the potential to enhance the fuel economy of road vehicles, by encouraging traffic to maintain economical patterns of flow. In urban areas this generally involves reducing congestion, so that vehicles can move more freely and the number of stops is cut. On principal routes, it may involve the use of speed limits to smooth traffic flows. A central element of traffic management theory is the design of junctions. By varying the priority given to two 'competing' traffic flows, delays at traffic junctions can be minimized. An urban traffic control system known as SCOOT (split cycle time and offset optimization technique) models traffic flow conditions and controls traffic signals accordingly, using a computer model of the road network. Over 40 SCOOT systems are currently in use throughout Britain.

Another example of traffic management technology that can enhance fuel economy is the 'Wolfsburg Wave' information system. The purpose of this is:

> ... to supply drivers in specially equipped vehicles with a steady flow of information about their relative position with respect to the green period of the next traffic signal, so that they are able to pass through the intersection while the signal stage is still 'green' by choosing the appropriate speed.
>
> Hoffman and Zimdahl (1988)

Tests on cars equipped with the system have shown that fuel savings of up to 6 per cent can be achieved, together with appreciable reductions in the number of stops. Further experimentation with fuel-efficiency technologies has been made possible under PROMETHEUS (Programme of a European Transport System with Highest Efficiency and Unprecedented Safety), undertaken by European car manufacturers.

Traffic calming is another form of traffic management, using techniques developed on the Continent. It is based on the principle of slowing down traffic in commercial or residential areas, in order to reduce its impact on pedestrians, cyclists and people using the street for other purposes. Measures include speed humps, chicanes, raised surfaces and 'gateways'. The relationship between traffic calming and fuel consumption is a complex one. Calming measures may tend to reduce fuel consumption and emissions. But this is generally the result of fewer vehicles using the 'calmed' streets, rather than any improvement in individual vehicle efficiency. In fact, a car may consume *more* fuel as a result of the obstacles that it has to negotiate in a traffic-calmed street. But the introduction of traffic calming has brought with it striking improvements in the local environment, as well as reductions in road accident casualties, and it is likely that these benefits outweigh any small penalty in fuel consumption and emissions that may exist.

Another factor affecting fuel consumption is the influence of *weather conditions*, and particularly temperature. Fuel consumption is some 10 to 15 per cent greater in winter than in summer, as a result of lower engine temperature, greater use of auxiliary equipment (heating and lighting) and increased aerodynamic drag arising from the higher density of the air. The contribution of weather-related equipment alone (heater, headlamps, wipers and rear window heater) is estimated to be 9 per cent (Martin and Shock, 1989).

Fuel economy in buses and coaches

It was shown earlier that typically only 18 per cent of the energy used by a car is associated with useful work, the remaining 82 per cent being accounted for by thermodynamic and mechanical losses. A similar figure applies to buses and coaches. Table 3.8 compares the specific energy consumption of different types of bus and coach operations. Double-decker buses and inter-urban coaches are the most economical in terms of energy consumption per passenger kilometre – as also indicated by Figure 3.5 earlier.

Table 3.8 Specific energy consumption of buses and coaches (1986)

Vehicle and journey type	Seating capacity	% vehicle occupancy	SEC (MJ per passenger km)
Minibus, suburban	15	25–50	0.8–1.6
Single-deck coach, suburban	33	25–50	0.6–1.2
Single-deck coach, motorway	50	25–50	0.5–1.0
Double-deck bus, city centre	75	50–75	0.3–0.4
Double-deck bus, suburban	75	25–50	0.5–0.9

Source Martin and Shock, 1989

Bus travel in towns and cities is usually characterized by low speeds, with frequent stopping and starting. In these conditions, aerodynamic losses are small compared with the energy used in accelerating the vehicle and in overcoming rolling resistance. Figure 3.13 shows how the useful work of an urban bus is divided between various aspects of operation. Aerodynamic drag is a minor energy 'sink', compared with acceleration and hill-climbing.

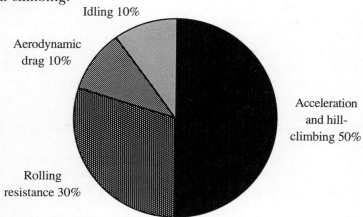

Note These figures refer to useful work, which constitutes typically 20 per cent of total energy consumption
Source Gotz, 1983

Figure 3.13 Use of energy in the operation of an urban bus

In rural and express services, stops are much less frequent and average speeds are greater. Aerodynamic drag becomes far more significant, accounting for 60 per cent of energy consumption at 100 kmh. The remaining 40 per cent is accounted for by rolling resistance alone, since acceleration and idling are not very significant areas of energy loss.

The extreme difference between these two types of operation means that buses cannot be designed for optimum fuel economy in any particular application. Instead, a compromise is made between the demands of a safe and easily accessible urban bus and a comfortable, aerodynamically shaped vehicle for use in inter-urban services.

As with private cars, factors that influence the fuel consumption of buses and coaches may be categorized into vehicle effects and operational effects. On board the vehicle itself, turbocharging, intercooling and fuel injection are some of the technologies which offer improvements in fuel economy and performance, as are thermostatic fans and radiator shutters. Further improvements can be achieved through other technologies such as low-viscosity oils and the use of precision cooling, whereby the engine coolant is applied only to those parts of the engine that become particularly hot. Another option available to bus operators is to purchase vehicles with large engines, for the sake of long service life, and then to 'derate and despeed' the engine in order to save fuel (Martin and Shock, 1989).

As with cars, fuel consumption in buses is also strongly influenced by the design of the transmission system. In cruising conditions, the availability of a suitable gear ratio to minimize engine speed can improve the fuel economy of a bus by up to 20 per cent. To the same end, continuously variable transmissions (CVTs) can be fitted to urban buses, with estimated fuel savings of 10 per cent (ibid).

The design of bus and coach engines has been strongly influenced by the introduction of European legislation on exhaust emissions. 'Euro 1', introduced in 1993, and 'Euro 2', scheduled for 1996, require significant cuts in the output of nitrogen oxides, carbon monoxide, hydrocarbons and particulate matter, compared with previous European standards. A variety of technologies are available for achieving the proposed Euro 2 limits, including intercooling, swirling the fuel/air mixture, and improved design of fuel injectors and piston ring and valve stem seals. According to Simpson (1993), 'manufacturers have done a commendable job in achieving Euro 1 without sustaining either the high penalties in fuel consumption or resorting to expensive and complicated bolt-on exhaust aftertreatments which were once thought to be an unavoidable cost of producing a green engine'. In particular, 'soot traps' – canisters introduced into the exhaust system to trap particulate emissions – will not necessarily be needed to achieve even the more stringent Euro 2 standards. Given the fuel economy penalty associated with exhaust traps, it is all the more encouraging that manufacturers have found ways of reducing both regulated pollutants and CO_2 through improved engine design rather than by resorting to exhaust treatments.

The physical weight of buses and coaches, measured per passenger seat, has generally increased in recent years. There are several reasons for this. The trend towards one-person operation in bus services has necessitated more powerful engines and additional automatic door equipment. For example, the traditional London Routemaster bus weighed 7.8 tonnes and could seat 72 people, whereas the Leyland Olympian buses that have largely replaced them weigh 10 tonnes and accommodate 68 people (Martin and Michaelis, 1992). Long-distance services have seen a widening of seats, and a general increase in the amount of equipment such as toilets, drink dispensers and entertainment systems. Vehicle weight could be reduced, and structural strength improved, by a switch to integral construction of the vehicle's body and chassis. It is estimated that in urban conditions a 10 per cent reduction in vehicle weight would lead to a 5 per cent improvement in fuel economy (ibid).

Large amounts of energy are lost by buses during braking, which can be very frequent in urban services where stops may be less than 500 m apart. Systems are available which allow a vehicle's momentum to be conserved during braking, rather than dissipated as waste heat in the braking system. Experimental designs have included flywheels and compressed air systems. These accumulate the kinetic energy of a vehicle during braking, store it whilst the vehicle is stationary, and then use the stored energy to accelerate the bus again. Flywheel systems comprise a rapidly spinning wheel which is connected to the vehicle transmission by means of a clutch. The wheel spins whilst the bus is stationary, and its energy is used to help get the bus moving again. In an alternative arrangement, the same function is performed by a large container filled with compressed air: the kinetic energy of the bus is used to compress the gas, and the system is operated in reverse to accelerate the bus. A test vehicle fitted with such a device, named 'Cumulo', was developed in Britain in the 1980s, but never used in commercial operations. The benefits of kinetic energy storage are dependent on the operating conditions, with maximum savings occurring in trips with frequent, short-duration stops. It has been estimated that fuel consumption could be reduced by up to 25 per cent using kinetic energy storage (White, 1986).

On the operational side, the way in which buses are managed and operated can have a significant effect on overall energy consumption. According to Martin and Shock (1989), 'driver education and increased awareness of techniques to improve fuel consumption appear to be the principal low-cost management activity'. Practices that might be encouraged among drivers include the avoidance of severe acceleration and braking, the moderation of speed, and switching off the engine whilst stationary.

For coaches used in inter-urban operations, the use of speed-limiting devices to prevent dangerous driving can also save fuel. By law, all coaches in Britain registered since 1974 must now be fitted with a device that limits their speed to 70 mph (113 kmh) or less. New legislation from the European Commission requires all new coaches in member states from 1994 to be fitted with speed governors calibrated to 100 kmh (62 mph), and this will eventually be extended to cover not just newly registered vehicles but all coaches in use.

In practice, these devices are in many cases inoperative, as a result of either malfunction or deliberate interference by frustrated drivers. Speed measurements published by the British government suggest that a large number of coaches are fitted with speed limiters that do not work, and the effect on fuel consumption is likely to be considerable.

Traffic congestion is another cause of increased fuel consumption in buses, as well as reduced service reliability, particularly in urban areas. A variety of 'bus priority' measures are available for improving the movement of buses, including bus-only lanes. The SCOOT traffic signal control system, which responds to traffic flows at a junction by optimizing the 'green time' of the signals, is available in a modified form whereby it can respond to an approaching bus by turning green. Used in conjunction with other bus priority measures, this technology can reduce fuel consumption by up to 10 per cent, both by improving the fuel economy of the bus and by cutting the duration of its journey.

Fuel economy in rail vehicles

Energy consumption in rail vehicles is determined principally by two factors: the mass of the train, which influences the energy required to accelerate it to its running speed, and the aerodynamic resistance encountered during operation. As with buses and coaches, the type of stock and the operating conditions will determine which of these is the more important. Mass is more critical in low-speed journeys with short station spacings, where stopping and starting represent a relatively large proportion of a train's operation. Meanwhile, aerodynamic drag is more important in high-speed or underground operations. A further energy demand comes from the heating and air conditioning of passenger compartments. Averaged over all seasons, between 11 and 15 per cent of energy consumption is for this purpose (Martin and Shock, 1989). Table 3.9 gives details of the primary energy consumption of different forms of rail operation in Britain.

The weight of rail stock directly affects not only the energy needed to accelerate it to running speed, but also, to a lesser extent, the rolling resistance. British Rail estimates that in urban services a 9 per cent reduction in vehicle weight offers an 8 per cent energy saving.

A major component of a passenger train's weight is the bogies. By building longer coaches, or allowing adjacent coaches to share a common bogie, the number of bogies per train can be reduced. The length of InterCity coaches used by British Rail has increased from 21 m in 1960s'

Table 3.9 Primary energy consumption of rail stock in Britain

Type	Energy consumption (GJ per 100 train km)
Diesel-powered	
High Speed Train	19.7
Main line locomotive	18.6
Diesel multiple unit	2.5
Electrically-powered	
Main line locomotive	23.9
Electric multiple unit	2.9
Underground train	9.7–12.2
Light rail	2.4–4.8
Urban electric train	12.5

Source Martin and Shock, 1989

stock to 23 m in the 1970s and 1980s. The Mk 5 stock for future trains is planned to use 26 m coaches, the maximum length that can be accommodated by the British loading gauge. Bogies themselves can be redesigned in lightweight form, with weight savings of over 50 per cent. The use of lightweight materials, such as aluminium, alloys and composites, offers weight reductions in passenger coaches, as do new construction techniques such as the fitting of body panels using adhesives rather than bolts and rivets (ibid).

New forms of diesel engine with greater efficiency offer fuel savings of around 5 per cent. But the need for increased noise control measures may impose a small penalty on fuel economy. Perhaps surprisingly, the use of rapid bursts of acceleration can save energy overall, because it allows longer periods of 'coasting', in which no energy is consumed. For example, an electric multiple unit can reduce its energy consumption by 17 per cent and its maximum speed by one-third, with no reduction in journey time, if the gear ratio is increased from 1:3.3 to 1:5.0. Microprocessors may be employed to optimize gear ratios, by controlling gearchanges, engine power, coasting and braking. Used this way, they can lead to energy savings of up to 30 per cent compared with conventional manually operated systems (ibid).

As with buses, there is scope in rail traction for conserving energy that would otherwise be lost as waste heat during braking. In conventional train braking systems, brake pads are used to convert the train's kinetic energy to heat, which is lost to the air. In an alternative arrangement, the motors may be operated in reverse to generate electric current, which creates heat in banks of resistors. But much of this waste can be avoided by converting the train's energy into electrical energy, which is fed back into the distribution network and reused – a principle known as *regenerative braking*. This is done by reversing the operation of the electric motors, and using them as generators driven by the train wheels. As the train slows down, electricity is generated in the motors and passed back into the overhead wires. Overall savings of around 15 per cent are possible using regenerative braking, and the benefits may be enhanced by the use of microprocessors. It is estimated that regenerative braking could save as much as 40 per cent of the energy used in suburban rail services in London, where stations are commonly spaced less than 2 km apart (Martin and Michaelis, 1992). The new Networker trains, recently brought into service in Britain, make use of regenerative braking as well as other other energy-saving technologies, and consume 24 per cent less electricity than their predecessors – despite increased speed and acceleration (Heaps, 1991).

It was stated earlier that aerodynamic drag is a function of frontal area and drag coefficient C_d, among other things. At full speed, a typical passenger train consumes approximately three-quarters of its energy in overcoming drag. When passing through tunnels, the aerodynamic drag force experienced by a train increases by between 100 and 200 per cent, and the fraction of energy consumed in overcoming air resistance rises to 90 per cent (Gawthorpe, 1983). The aerodynamic design of passenger trains is therefore crucially important in determining energy consumption. The principal contributors to aerodynamic drag in trains are the bogies and the surfaces of the rolling stock. Interestingly, the front of the train contributes only around 5 per cent to overall wind resistance.

In the last 15 to 20 years the aerodynamic design of passenger trains in Britain has improved by around 40 per cent. Much of this has been the result of better fairings, smaller gaps between cars and the elimination of opening windows and roof-mounted ventilators from many trains. However, the latter are associated with the introduction of air conditioning, which imposes an additional energy demand. Most of the potential for further improvements in aerodynamics lies in locating extraneous equipment beneath the floor of the coach and surrounding it with fairings. Similar practices can be applied to the bogies, although this makes access for maintenance more difficult. Similarly, the design of the pantograph (the mechanism that picks up power from overhead cables) can be modified to reduce its drag coefficient (Martin and Shock, 1989).

Modern express trains have a drag force typically 20 per cent less than that of 1960s' and 1970s' stock. It is estimated that further reductions of 10 to 15 per cent are possible without radical design changes. In services where operating speeds are lower, such as those on local lines, drag reductions of up to 40 per cent are possible in individual trains. But because of the low speeds, the benefit is translated into a modest 5 per cent reduction in energy consumption overall (ibid).

A growing proportion of rail services is electrified, rather than powered by diesel locomotives. Two major benefits are associated with electrification. Firstly, electric traction is cheaper than diesel: Hamer (1979) estimates that when capital investment and maintenance costs are taken into account, electricity is some 30 per cent cheaper, partly because of lower energy costs. In terms of delivered energy, electric trains are more efficient than diesel (though when the losses associated with the generation and distribution of electricity are considered, there is little difference between the two). Electric engines are more reliable than diesel locomotives, as a result of their relative simplicity. This means that fewer trains are out of service at any time, and maintenance costs are between 50 and 65 per cent less than those of diesel stock. Additionally, electric trains have better performance characteristics than diesel. An electric engine has a power-to-weight ratio some 50 per cent greater than an equivalent diesel locomotive. It can also withstand short bursts of power in excess of its continuous rating, giving impressive acceleration and hill-climbing characteristics (Martin and Shock, 1989).

The main drawback associated with electrification is that it generally increases peak electricity demand. In countries like Britain, this means that the additional electricity would be generated by relatively inefficient coal-burning power stations. In other countries, where coal is used less intensively, increasing electricity demand could have less of an environmental penalty.

Recent electrification projects, such as Britain's East Coast Main Line, have been justified principally on economic grounds. But routes with little traffic cannot make such a justification. Potter (1990) summarizes the situation as follows:

> Electrification on British Rail is undertaken within the broadly commercial remit of the business sectors. Recent projects have majored on the cost-cutting potential of electrification and (to a lesser extent) its proven traffic-generating potential. However, with replacement of 1950s modernisation diesels approaching completion, further major electrification appears unlikely.

In terms of primary energy demand, and emissions of CO_2, there is currently little or no advantage in electrification when compared with diesel traction. However, if electricity generation and distribution were to become more efficient, as seems likely, or if renewable forms of electricity generation were to become more widespread, electric traction could develop a significant advantage over diesel in terms of both energy efficiency and CO_2 emissions. In addition, the traffic generation effects should not be overlooked: if electrification has the effect of attracting travellers to rail from less energy-efficient modes such as car or airline, there would be a further reduction in CO_2 emissions. Martin and Michaelis (1992) summarise the position thus:

> Even if electric trains do have higher greenhouse emissions than diesels, as in the UK at present, this does not necessarily mean that emissions per passenger kilometre are higher. Rail operators have not electrified their track out of concern for global warming. Electric trains are able to operate faster and more reliably than diesel trains. They therefore attract more passengers and may have higher occupancies.

Energy savings are also associated with the elimination of gradients and curves from rail lines, leading to reduced braking and acceleration. But realigning routes is an expensive operation, and possible only in a small number of cases. More important is the easing of very low speed limits (for example 60 kmh curves or crossings) on high-speed lines, and other local measures. Such work is often undertaken as part of wider engineering projects, such as electrification and resignalling.

Finally, considerable energy losses are incurred when rail traffic is required to slow or stop whilst travelling along track that is being repaired. Savings in both time and energy could be made if the maintenance work were to be modified so that speed restrictions were fewer and lesser. An example of this can be found in the new Dynamic Track Stabilizers, which allow new track to be used immediately rather than after a 'bedding in' period (British Rail, 1991). Previously, speed limits as low as 30 kmh were imposed for up to two weeks after new track was laid.

Fuel economy in aircraft

As mentioned earlier, the specific energy consumption (SEC) of air travel has declined substantially in recent years. This improvement has been characterized by a trend towards larger aircraft, which have the dual advantage of both providing greater energy efficiency and accommodating a larger number of passengers.

Long flights are generally more fuel efficient than short ones, partly because short flights tend to be made by smaller, less fuel-efficient aircraft. But there is a more fundamental basis for the difference. The 'cruise' phase of a flight is generally the most economical in terms of fuel use, whilst climbing consumes a large amount of energy. In a typical 900 km flight, the initial ascent consumes as much fuel as the entire cruise phase (Martin and Shock, 1989). In short flights, such as those between UK destinations, ascent and descent account for a large proportion of the journey. Thus for any type of aircraft, average fuel economy improves as the length of flight increases. The improvement tends to level off for distances greater than 1,000 km.

Martin and Shock (1989) have described in detail the aspects of aircraft design that can contribute to improved fuel consumption characteristics. Progress in computer-aided design has allowed better flow modelling to be performed, and there has also been some experimentation with laminar flows and grooved skins. At present, aluminium alloys are the principal material in the structure of aircraft, and are likely to remain so. However, alloys with superior physical properties are being developed, while composite materials such as glass fibre, kevlar and carbon fibre are likely to become more common in airframe components.

Most medium and large aircraft are powered by turbofan engines, whilst smaller aircraft, used in shorter trips, tend to have turboprop engines. Both fuel consumption and maximum thrust have increased significantly since the early days of passenger jets. In addition, computers are being used increasingly to control aircraft. For example, the

A320 Airbus is the first commercial aircraft to be equipped with 'fly by wire' controls, with computers providing an interface between the pilot and the control surfaces. The introduction of computerized controls offers significant reductions in weight and fuel consumption. Computers may also be used to devise flight patterns in such a way that fuel consumption can be minimized. Technologies such as satellite navigation, microwave landing systems and advanced weather forecasting can also reduce the amount of fuel consumed by aircraft during a typical flight (ibid).

Summary

In summary, vehicle fuel consumption, alongside occupancy, has a direct effect on emissions of CO_2. This chapter has described the many factors that influence the fuel consumption of different modes, broadly divided into vehicle effects and operational effects.

In general, vehicles return a considerably lower level of fuel economy than the maximum that is technologically possible. This is particularly noticeable in private cars. In part this is because it takes time for new, more efficient designs to find their way into the vehicle stock. But a more important reason for the discrepancy is that other characteristics, such as comfort and performance, tend to be traded off against fuel economy. The point at which the trade-off occurs is determined crucially by the price of fuel. Experience from the car markets of a number of European countries suggests that long-term fuel price has a positive influence on the average fuel economy of the national car stock.

Operational effects too can have a considerable influence on fuel consumption. In the case of cars, this includes driver behaviour and measures to reduce uneconomical driving conditions such as congestion and cold running. In other modes, particularly rail and air, new operating systems incorporating microprocessor controls may be employed to ensure that vehicles are operated in the most economical way possible. In public transport modes, the fuel costs, taken as a proportion of total operating costs, are high compared with those for cars, and there are significant commercial gains to be made from improving fuel economy.

There is a conspicuous gap between real-life fuel consumption and what is technologically possible. With fuel at current prices, consumers tend to trade off fuel economy in favour of other benefits. The wide gulf that lies between actual and potential fuel economy represents a huge, untapped resource of energy efficiency which may be exploited as part of a strategy to curb emissions of CO_2.

Notes

1. Excluding nuclear power stations.
2. 'Petroleum' is a general term for oil products, including motor spirit, diesel and aviation fuel.
3. The percentage is higher if subsidiary activities, such as vehicle manufacture and infrastructure provision, are included.
4. 'Load factor' is a term describing the fraction of places occupied.
5. Gasoline consumption is used as an indicator of the demand for car travel. It should be remembered, however, that other factors besides travel demand, such as vehicle fuel economy, have an influence on the amount of gasoline consumed.
6. In first and second place are iron and steel (252 to 284 PJ) and organic chemicals (132 to 148 PJ).
7. Methane is normally removed from the atmosphere through a reaction with OH, so if the abundance of OH becomes less, methane proliferates. High levels of methane in the atmosphere can also lead to an increased concentration of water vapour, itself a greenhouse gas, when the methane is oxidized (Ramanathan, 1988).
8. When considering emissions from aircraft, it is worth noting the significance of altitude. According to the European Commission (1992) the impact of greenhouse gases such as N_2O is considerably greater in the middle to upper troposphere than at ground level.
9. The pollution problem associated with cold-starting in catalyst-equipped vehicles can, however, be partially solved by preheating the catalyst, or the whole engine. Details of this technique appear later in the chapter.
10. Until recently the most economical car available on the British market, and probably world wide, was the Daihatsu Charade Diesel. But little mention was made of this feature in Daihatsu's advertising campaigns.

    Addressing the motor industry on the eve of the British government's first environmental White Paper in 1990, the then environment secretary Chris Patten said: 'It cannot be sensible to develop and advertise cars that can travel well over 100 miles per hour, with rocket-like acceleration, whose performance potential can be exploited only by disregard for the environment and by breaking speed limits' (Clover, 1990).
11. It is also worth noting that emissions of carbon monoxide, hydrocarbons and nitrogen oxides from a diesel engine are lower than those from a petrol engine fitted with a three-way catalytic converter, even after correcting for the lower fuel consumption of a diesel (Cragg, 1992).
12. This is because the cyclical compression of tyres in motion is not perfectly elastic, and some energy is converted to heat.
13. Professor Stephen Salter, who became famous for his wave-power 'ducks' in the 1970s, has more recently used his engineering expertise to develop a hydraulic CVT system that could be used in road vehicles (Pendrous, 1990).
14. Although diesels are generally more reliable than petrol engines, poor maintenance tends to be less apparent to the driver – though no less polluting – in the case of a diesel engine (Cragg, 1992).
15. Researchers at Volkswagen in Germany have developed a device which automatically switches off the engine while the vehicle is stationary – for example, at traffic lights or in queues. Significant fuel savings have been reported using a VW Golf fitted with the system.

4

'Business as Usual'
Carbon Dioxide Emissions

> Transport, and road transport in particular, will present
> insoluble problems in terms of pollution, energy consumption
> and congestion unless there are policy changes.
>
> The European Parliament

In order to assess the scale of the 'greenhouse' threat posed by personal
travel, it is necessary to estimate how emissions of carbon dioxide are
likely to change in the future. Such a projection provides a useful
reference point, or baseline, against which alternative strategies can then
be assessed.

The use of 'scenarios' is well established as a tool for policy research.
By employing them it is possible to examine different possible futures,
and to assess the effect of different forms of intervention. Many readers
may be tempted to throw up their hands in despair at the mention of the
's'-word as, in the last few years, the scenario approach has been used
extensively for examining future levels of emissions, particularly CO_2,
under different sets of assumptions. However, the key advantage of such
an approach lies in the distinction between a scenario and a forecast.
Using a set of related scenarios, a number of possible futures can be
formulated, often variants of one another, and their consequences
examined, with no need to speculate as to which is the 'most likely'.

To establish how CO_2 emissions from personal travel are likely to
change in the absence of any intervention, it is necessary to construct a
'business as usual' scenario. This incorporates all the relevant policies
that are currently in place, and extrapolates them to a certain point in the
future. For the purposes of this book, the year 2025 is adopted as the
time horizon. 'Business as usual' is defined as the progression of events
that will take place as a continuation of present policies and trends,
assuming that no policy changes will take place other than those

currently under development. For example, energy efficiency might be expected to improve as a result of market mechanisms, but no more.

It is most important to bear in mind that 'business as usual' does not imply the absence of any action, or 'policy neutrality'. On the contrary, many transport policies currently in place have a direct effect on the volume and nature of travel undertaken, and the resulting emissions of CO_2. For example, the level of bus and rail transport fares relative to the cost of driving has a major influence on the modal share held by public transport. Likewise, land-use policies that allow unrestricted development on the urban fringe tend to encourage the use of cars for everyday trips to work, school or shops. 'Business as usual' assumes a continuation of policies such as these.

The SPACE model

In order to examine personal travel demand under different future scenarios, a computer model has been created known as SPACE – Scenario Projections of Aggregate Carbon Emissions. SPACE is a spreadsheet model which describes personal travel in fine detail up to the year 2025, in five-yearly intervals. Combining information on trip lengths, trip frequencies, fuel economy and other factors, the model provides a five-yearly chronicle of CO_2 emissions from each mode of travel.

SPACE is a model of personal travel in Britain, based on detailed travel survey information. Accordingly, the results produced by the model relate to CO_2 emissions from Britain alone. However, many of the principles established by the modelling exercise are applicable, in varying degree, to other developed countries, particularly those in Europe with similar geographical and socio-political characteristics.

The model has as its inputs various aspects of personal travel demand. Its output takes the form of an annual quantity of carbon dioxide resulting from personal travel in Britain as a whole. The base year for all projections is 1988. Using data from the two most recent National Travel Surveys, the SPACE model breaks down travel demand into the following variables:

- Household car ownership
- Journey purpose
- Journey length
- Journey frequency
- Choice of mode
- Residential area (urban, intermediate or rural).

The data take the form of average journey lengths (in miles), and average journey frequencies (in trips per year) for each of the following household categories:[1]

- Rural with cars
- Rural without cars

- Intermediate with cars
- Intermediate without cars
- Urban with cars
- Urban without cars.

Within these user categories, the journeys undertaken are subdivided according to trip purpose, namely:

- Work
- Shopping and personal
- Social and entertainments
- Holiday and other.

This analysis was performed for five modes of travel, namely car, rail, bus, walk and cycle, in addition to more simplified calculations for air, motorcycle and light rail travel.

Population and car ownership

Population is a key determinant of passenger transport demand. The SPACE model considers the proportion of the total population which

- lives in car-owning households, and
- lives in rural, intermediate and urban areas.

The total population of Britain, like that of most developed countries, is expected to grow very slowly, by a little more than 0.1 per cent per annum. Superimposed on this trend, however, the British government has forecast a vigorous growth in car ownership, rising from 310 cars per 1,000 population (cpt) in 1986 to between 523 and 582 cpt in 2025 – a little below the level of car ownership currently found in the USA.

It is assumed for Scenario 1 – 'business as usual' – that, in all three types of geographical area, the proportion of people living in households with one or more cars will rise considerably. In rural areas, 95 per cent of people are expected to live in car-owning households by the year 2025, while the figures for intermediate and urban areas are 90 and 85 per cent respectively (see Table 4.1).

Table 4.1 Estimated 'business as usual' growth in Britain's car ownership

	% of people in car-owning households	
	1988	*2025*
Rural areas	87.2	95.0
Intermediate areas	78.3	90.0
Urban areas	69.7	85.0

Note 'Car-owning' is defined by households with one or more cars

To estimate the geographical distribution of population, it is helpful to examine past trends. For many years Britain's major cities have been losing people at a significant rate – between 10 and 20 people per 1,000 population per year. Other metropolitan areas, as well as smaller cities, have had a lower shrinkage rate, while industrial areas, new towns, resorts and other areas have been growing. This trend has tended to level off since the mid-1970s, with rates of growth or decline becoming more modest (Champion, 1987a).

The change in population distribution observed in the period 1985–9 is likely to continue for some time. The main growth is expected to continue to be in rural and intermediate areas, with less change in urban population (ibid). For Scenario 1 this trend has been adopted (see Table 4.2).

Table 4.2 Population trends projected in Scenario 1

	Change in population (% per annum)
Urban areas	−0.08
Intermediate areas	+0.58
Rural areas	+1.10

Source OPCS, 1990

Transport demand

According to Dr Phil Goodwin of Oxford University, forecasts are 'ubiquitous, embarrassing, usually wrong, rarely disprovable, often useful – and manifestly necessary' (Goodwin, 1993). Nowhere is this criticism more appropriate than in the case of the National Road Traffic Forecasts (NRTF) for Great Britain, published in 1989, which contain revised predictions of road traffic up to the year 2025. The NRTF predict traffic levels which are considerably higher than any previous government estimate. Road traffic volume is forecast to grow linearly, reaching between 183 and 242 per cent of its 1988 value by the year 2025 (see Table 4.3). The main determinant of traffic volume is considered to be economic growth.

Table 4.3 Summary of the 1989 Road Traffic Forecasts for Britain

	Growth in vehicle kilometres 1988–2025 (%)	
	Low	*High*
Cars	82	134
Light goods vehicles	101	215
Heavy goods vehicles	67	141
Buses and coaches	0	0
All traffic	**83**	**142**

Source Department of Transport, 1989a

Not surprisingly, concern has been expressed by many that the growth rates implicit in the 1989 NRTF are unrealistically high. The consequences for greenhouse emissions are one cause of this concern, whilst many have pointed to the social and land-use implications of such a large-scale expansion in road traffic.[2] The rapid rate of traffic growth projected by the 1989 NRTF has prompted criticism of the forecasting techniques employed by the DoT. The most common objection is the use of economic growth (GDP) as the prime determinant of traffic growth, to the virtual exclusion of other factors such as environmental and political pressures. According to Cragg (1992), when the NRTF were published:

> ...it seemed that the entire rationale of the forecasting exercise was to decide whether to build more roads now or later. The Department did not seem to consider any alternative.
> It did not take critics to point out the circularity hidden in the forecasts. The fact was that the forecasts of vehicle use growth were devised by the Department of Transport as a means to 'appraise trunk road improvements' but had within them the implicit assumption that these road improvements would take place.

The 1989 forecasts brought into relief the need for 'saturation' effects to be taken into account, whereby physical limits are placed on the amount of traffic that can exist. The DoT assumed that saturation – in terms of both car ownership and car mileage – would not occur during the period of the 1989 forecasts. The Royal Town Planning Institute (1991) has expressed the view that the historical growth in road traffic has largely been absorbed by existing road capacity, which is now unable to cope with further increases in vehicle numbers:

> ...it may no longer be reasonable to assume that the physical fabric, or the pattern of uses and trips within the fabric, can continue to absorb change. If traffic increases over recent decades have been largely absorbed within the existing fabric, and that capacity largely used up in many areas, further growth in traffic demand may create accelerating pressures for change in urban form, even when the actual growth in demand may be decelerating.

These concerns are related to the observation that most of the government's roadbuilding programme consists of trunk roads beyond the limits of urban areas, and will encourage greater numbers of vehicles to enter already-congested towns and cities.

Although the Department of Transport still officially expects road traffic to grow at the rates forecast in the 1989 NRTF, it has acknowledged that traffic growth in some areas may be constrained by limited road capacity. In its White Paper *Trunk Roads in England in the 1990s* it expresses the view that 'there will be cases where, on economic or environmental grounds, it is neither practicable nor desirable to meet the demand by roadbuilding, for example in city centres'.

Further doubt has been cast on the reliability of the 1989 NRTF by the effect of the recent recession. Between 1989 and 1992, a remarkable 'plateau' in traffic levels appeared, contrasting sharply with the rapid growth of the 1980s. In 1992, road traffic actually dipped below the Department of Transport's 'low' forecast for that year (Hughes, 1993).

Cars

The 1989 NRTF assume an annual growth of between 2.2 and 3.6 per cent in car traffic. Whilst it is by no means clear what rate of annual growth is sustainable for the period to 2025, it is assumed for the purposes of the SPACE model that the upper forecast will be a more likely outcome than the lower, in a 'business as usual' world. The car sector model, based on the travel patterns revealed in the 1985/86 National Travel Surveys, was therefore scaled in such a way as to produce an overall traffic volume marginally below the 'high' forecast of the 1989 NRTF.

Translating car kilometres to passenger kilometres requires assumptions to be made concerning vehicle occupancy. Chapter 2 showed that car occupancy has historically declined as the number of cars in use has increased. This trend is projected forward in Scenario 1 in a linear fashion, with car occupancy expected to decline in all areas by 0.1 per cent per annum. The lowest car occupancy rate in 2025 is expected to be in rural areas, with an average figure of 1.47 persons per car. Occupancies are expected to remain higher in urban areas, in accordance with historical trends revealed in recent National Travel Surveys.

Buses and coaches

For bus and coach travel, the 1989 NRTF predicts a zero growth in vehicle kilometres up to the year 2025.[3] But this does not, however, imply a zero growth in passenger kilometres. The trend towards smaller vehicles that has taken place in recent years, particularly after the deregulation of bus services in 1986, has meant that more vehicles are required to carry the same number of passengers. Furthermore, the population of Great Britain is expected to be some 5 per cent greater in 2025 than it is at present (see above).

The 'zero growth' prediction of the 1989 NRTF for bus and coach travel therefore implies that bus travel *per person* will actually decline. It is therefore assumed in Scenario 1 that, while bus kilometres remain unchanged, bus passenger kilometres and occupancy will both fall. The decline in patronage that has occurred since deregulation in 1986 is particularly sharp, and it is expected that it will bottom out to a level of demand representing a 'core market' of bus users.

Rail

For the rail sector, Scenario 1 assumes a continuation of the annual growth in passenger mileage that has taken place in recent years.

Additionally, urban light rail systems are expected to make an increasingly significant contribution. In 1993, urban rail schemes were operating in London, Glasgow, Manchester, Newcastle and Blackpool, with a project nearing completion in Sheffield. Other city-wide LRT proposals were being developed. In view of the interest shown in light rail, 'business as usual' assumes a 50 per cent increase in LRT passenger mileage by the year 2025.

In the decade from 1979 to 1989, the average number of passengers carried by British Rail per train has varied little. In 1981 it dropped to 92, whilst in 1988 it reached 101. Currently BR is trying to expand passenger capacity without extending the vehicle stock, and increased seating capacity is an active policy. However, it should be noted that expanding the seating capacity of carriages serves largely to provide seating for passengers who would previously have had to stand, rather than enabling additional passengers to use the train. For this reason the growth in average occupancy on BR trains is expected to be modest, reaching 111 persons per train by 2025.

The opposite effect is expected to take place in light rail transit. Traditionally the London Underground has accounted for most of the non-BR rail mileage in Britain. Loading factors here have shown an increasing trend throughout the 1980s as the demand for travel has risen. In 1987 the average number of people carried by a London Underground train reached a peak of 129. However, other urban rail schemes, whether operational or under construction, will not achieve such high occupancies as those typical on the London Underground. The Tyne and Wear Metro in Newcastle, for example, has smaller trains than London Underground – 84 seats per train and an average occupancy of 30 people (Withrington, 1990). As new LRT schemes become operational, the proportion of mileage travelled on the London Underground will therefore diminish, and the average occupancy for all urban rail travel will be expected to decrease. It is estimated that the average occupancy of a light rail train in 2025 will have fallen to 70.

Air travel

For domestic air travel, demand is expected to continue to increase. But this growth will be subject to constraints, particularly on runway availability and airport terminal capacity. One consequence of these limitations is likely to be an increase in the average seating capacity of aircraft, so that runways and airspace are used more efficiently.

Martin and Shock (1989) have taken account of these constraints and derived projections for air travel up to the year 2010. The first of their scenarios envisages a high rate of economic growth, a low rate of increase in fuel prices, and 'slow technical change'. This set of air traffic projections is adopted for the purposes of this book to represent 'business as usual'. The demand for domestic air travel rises from 4.7 billion passenger kilometres in 1988 to 14.0 billion in 2025.

As explained above, the increasing pressure on aircraft infrastructure is likely to produce an increase in the average seating capacity of aircraft. Load factors, however, will not necessarily be affected. Scheduled service load factors fell from 68 per cent in 1963 to 61 per cent in 1986. Although load factors on international and non-scheduled services tend to be higher, it is not expected that those on domestic flights will rise appreciably in the future. The Civil Aviation Authority expects this figure to remain constant at around 60 per cent. It is therefore expected that aircraft load factors for domestic services will remain constant at 60 per cent up to the year 2025, regardless of changes in seating capacity (ibid).

Other

For motorcyles, petroleum consumption and CO_2 emissions are comparatively small, and not very significant in an analysis of transport energy consumption. 'Business as usual' envisages a 25 per cent growth in motorcycle traffic, by the year 2025.

No forecasts are available on the likely long-term trend in cycling and walking, and estimates for the purposes of Scenario 1 are instead based on past trends. 'Business as usual' is expected to involve a 30 per cent increase in cycle mileage between 1988 and 2025, reflecting a continuation of the current revival of interest in cycling as a travel mode. Walking is expected to decline by 25 per cent in the same period, as journeys are transferred to cycling and motorized transport.

Energy consumption

Having established 'business as usual' estimates of travel demand for each mode up to the year 2025, it is necessary to consider the energy demand resulting from this scenario. To do this, assumptions must be made for each mode concerning (i) fuel economy, and (ii) the nature of fuels used. The approach adopted here will consider *primary energy* consumption; that is, the total energy demand of a transport mode including the energy costs of producing, storing and delivering the fuel. The importance of using primary, rather than delivered, energy as a measure was outlined in Chapter 3.

Cars

As discussed in the previous chapter, the fuel economy of road vehicles is a crucial determinant of overall energy consumption and CO_2 emissions arising from personal travel. Historically, car fuel economy has shown a slight improvement from year to year as automotive technology has enhanced the efficiency of new vehicles. However, fuel prices have generally not been sufficiently high to stimulate great interest in fuel economy, and the improvements in technology have been used primarily to increase the power of a given engine capacity, rather than in securing better average fuel economy. In Scenario 1, it is assumed that market mechanisms, coupled with technological advances, will continue to yield modest improvements in average car fuel economy.

Of equal importance is the distribution of the national car stock between different categories of *engine size*. As engine technology advances, a given power output can be delivered by progressively smaller engines. This implies that the fraction of cars in the largest engine size category would decline in favour of smaller engine sizes. However, in the absence of regulations or market mechanisms favouring fuel economy, there is likely to be little incentive for consumers to purchase more economical vehicles, and the dominant effect will be a 'trading up' to more powerful models. The proportion of cars in the smallest engine size category is not, therefore, expected to grow to the extent that technological progress might allow.

This effect can be observed in the historical figures for car fuel consumption. Figure 3.10 has shown that car fuel economy in Britain improved only marginally in the decade to 1987, because developments in vehicle efficiency were applied to the production of more powerful cars rather than more economical ones. Figure 4.1 illustrates the engine size distribution of cars up to the year 2025 in Scenario 1, 'business as usual'.

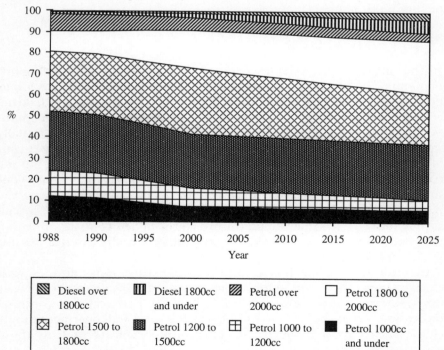

Figure 4.1 Engine size distribution of cars in Scenario 1
(fractions of total stock)

Consideration should also be given to the average fuel economy associated with each engine size category. Historically, technical advances have produced a modest annual improvement in the fuel economy of a given engine capacity, as explained in Chapter 3. 'Business as usual' assumes a continuing, but modest, improvement in the fuel economy of each engine capacity range.

Buses and coaches

For buses and coaches, the 'business as usual' trend in fuel economy is expected to be influenced primarily by the changing nature of the vehicles in use. As described earlier, the late 1980s have seen a trend towards smaller buses, which typically consume less fuel than traditional single or double-decker buses (though their smaller seating capacity means that their fuel consumption *per seat kilometre* is generally higher than that of larger vehicles). Scenario 1 assumes a continuation of the trend towards smaller vehicles, and thus a progressive improvement in the average fuel economy of buses. Coaches, on the other hand, are not expected to undergo changes in body size. Their fuel consumption is not expected to improve other than by a small amount resulting from technical refinements.

Rail

In the rail sector, there is less scope for fuel economy improvements. Economic performance already requires that fuel costs be kept as low as possible. Nevertheless, as new stock is introduced, it is expected that fuel economy will improve, as a result of new technologies such as regenerative braking, reduced vehicle weight and better aerodynamics. Equally significant is the degree of electrification. Currently around 50 per cent of British Rail's train kilometres are covered by electric traction, but the consequences of further route electrification for overall energy consumption are considerable. Since electric trains are powered by the national electricity supply, the sources of energy used by the electricity supply industry to generate power are a crucial determinant of overall CO_2 emissions. (The benefits of electrification were discussed at greater length in Chapter 3.)

In 'business as usual', it is assumed that electrification continues, with electric traction accounting for 80 per cent of train kilometres by the year 2025. Energy consumption per train kilometre is expected to continue increasing, particularly for light rail, which is expected to be increasingly characterized by lightweight, street-running tramcars such as those operating in Manchester.

With a growing proportion of rail travel electrified, the nature of the electricity supply industry will be increasingly important in evaluating emissions of CO_2. Most of Britain's electricity is currently derived from steam turbines powered by fossil fuels, principally coal. The efficiency with which fossil fuels are used to generate electricity is notoriously low: typically 70 per cent of the fossil fuel energy is lost during generation and transmission, mainly as a result of thermodynamic constraints (Davies, 1991).

It appears unlikely that nuclear power will be expanded on the scale once envisaged by the industry. The prohibitive cost of decommissioning nuclear stations, together with persistent concerns over the risk of a Chernobyl-scale accident, appear to have dealt a near-fatal blow to the world nuclear industry. Instead, the 'dash for gas' is creating a new generation of cleaner, and relatively efficient, gas-fired power stations. Meanwhile the exploitation of renewable energy sources, including wind, wave and tidal power, as well as energy from domestic refuse incineration, is increasing. Looking to the future, 'business as usual' is difficult to judge for Britain's electricity supply industry, in the light of uncertainties over plant characteristics and future demand.

For Scenario 1, the market shares held by different fuels are based on scenarios produced by the Department of Energy (1990). It is assumed that the expansion of gas as a fuel for power stations will continue, reaching a market share of 37 per cent by the year 2025. Coal is expected to decline to 45 per cent of the market, compared with 65 per cent in 1988; while nuclear and renewable sources, which for the purposes of the model are assumed to have negligible CO_2 emissions, fall to 15 per cent of the electricity supply industry, chiefly as a result of nuclear power stations reaching the end of their useful lives. Steam plant efficiency is expected to improve as new plant are built, although thermodynamic constraints will limit the scale of this improvement.

Air travel

In the air travel sector, it is expected that the increasing pressure on airport capacity caused by rising demand is likely to result in the use of larger aircraft, whose greater seating capacity will result in a more efficient use of airspace. The SEC of aircraft used for journeys within the UK is projected to increase by 13 per cent over the period 1988–2025, while seating capacity increases by a greater margin. The net result, for the purposes of Scenario 1, is therefore expected to be a modest improvement in fuel economy per passenger kilometre.

Other

Similarly, motorcycle fuel economy is expected to continue the improvement that has taken place in recent years, with average fuel consumption falling from 3.52 litres per 100 km in 1990 to 2.95 litres per 100 km in 2025. European legislation aimed at restricting the horsepower of large motorcycles, if approved, will contribute to this improvement.

'Business as usual' CO_2 emissions

The SPACE model was used to draw together all the projections outlined above into a 'business as usual' scenario. The outcome, in terms of passenger transport's total CO_2 emissions, is an overall increase of 80 per cent, from 87 million tonnes in 1990 to 157 million tonnes in 2025. This growth is illustrated in Figure 4.2.

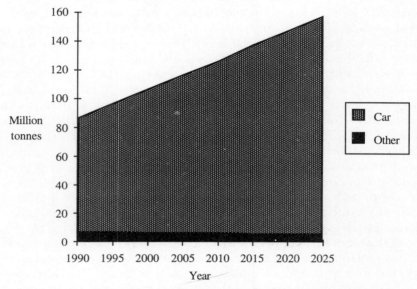

Figure 4.2 Total emissions of CO_2 from personal travel in Scenario 1

The growth in CO_2 emissions projected by the SPACE model compares well with forecasts made by the British government. Assuming a medium level of economic growth, it has estimated that CO_2 emissions from transport (including both passenger and freight operations) will increase by between 64 and 70 per cent by the year 2020 (Department of Trade and Industry, 1992).

The British government's policy on greenhouse emissions, in line with that of the European Community as a whole, is to stabilize CO_2 emissions at their 1990 levels by the year 2000. By contrast, Scenario 1 indicates that in the period 1990 to 2000, CO_2 emissions from passenger transport will increase by 23 per cent, from 87 to 107 million tonnes. This discrepancy raises important questions about whether, and how, the year 2000 target is likely to be achieved.

Chapter 3 has shown that transport operations currently contribute around 24 per cent to total CO_2 emissions in Britain. Of this, passenger transport contributes around two-thirds, or 16 per cent of national emissions. Assuming that passenger transport develops in the 'business as usual' manner outlined in this chapter, and that the nation as a whole simultaneously achieves its target of stabilizing emissions at 1990 levels by the year 2000, then it follows that all other sectors of the economy – industry, domestic, agriculture, commerce and freight transport – would between them need to achieve a 5 per cent *reduction* in CO_2 emissions by 2000, if the national target were to be met.

Having glimpsed 'business as usual' in the passenger transport sector, we are faced with a choice. Is it reasonable to expect the rest of the economy to make significant cuts in emissions while passenger transport continues to increase its own emissions? Or should personal travel be required to curb its projected growth in emissions along with all the other sectors? The approach adopted in this book is the latter. Even if transport can justify a less stringent CO_2 reduction target than that applied to other sectors, it should nevertheless play its part in securing a national reduction in greenhouse emissions. The following chapter explores some of the policy tools that might be drawn upon in order to dampen the projected growth in CO_2 emissions from personal travel.

Notes

1. 'Rural' refers to households in settlements of fewer than 3,000 inhabitants; 'intermediate' refers to those in settlements between 3,000 and 100,000 inhabitants; and 'urban' refers to households within settlements with more than 100,000 inhabitants.
2. See, for example, Adams (1990) and the Royal Town Planning Institute (1991).
3. The Royal Town Planning Institute (1991) suggests that greater attention should have been given to bus demand in the compilation of the NRTF: 'Significant transfers from car to bus might only be achieved by policy, but the land use dispersal associated with increased car use would be likely to be lengthening bus trips, or causing former walk trips to become bus trips.'

5

Policy Tools

A completely free market economy treats many environmental resources as if they have zero prices. Like any free goods,
these resources are then abused and overused. It is therefore
important to 'price the environment' as far as we can.

Professor David Pearce, University College London

Sharing the greenhouse burden

The previous chapter showed that a continuation of present policies,
with no action taken to curb greenhouse emissions, is expected to
produce an increase of 80 per cent in CO_2 emissions from personal travel
in Britain by the year 2025. By contrast, the current policy of the British
government on CO_2 emissions is to stabilize output at the 1990 level by
the year 2000.

A vital question at this point is, to what extent different energy sectors
(domestic, industrial, transport and so on) should curb their emissions in
order to contribute to a particular national target. The cost of stabilizing
greenhouse emissions varies between different sectors, so it is reasonable
that some method of allocation should be devised in order to minimize
the overall cost.

On this question the House of Commons Energy Committee (1991)
holds the view that:

A large increase in the transport sector's CO_2 emissions balanced by
disproportionately heavy reductions in other sectors would not be a
rational way of achieving the emissions target, and the Department of
Transport should draw up a comprehensive policy ... to try to prevent this
occurring.

No such policy statement has yet been produced by the Department of
Transport in response to this recommendation. Instead, the government

acknowledges somewhat generally 'that the transport sector will have to play its part in meeting any greenhouse gas targets' (ibid). At the time of writing, the government had committed itself to drawing up targets for reducing CO_2 emissions in individual sectors of the economy, but given no indication of whether transport would be treated more leniently than other sectors.

The German government, by contrast, has performed the necessary allocation exercise and produced figures for the cuts in emissions that various sectors might need to make. In order to achieve a 25 per cent cut in overall CO_2 emissions, some sectors would be required to make cuts as large as 39 per cent, whilst transport will be required to achieve just 9 per cent (Warren, 1990).

Although no such detailed strategy has been put forward for Britain, or most other developed countries, the German plan for sector-specific reductions in CO_2 emissions provides a good model for other nations to follow. For example, if a government such as Britain's were to adopt a target for reducing total CO_2 emissions by 60 per cent, as recommended by the IPCC (Houghton et al, 1990), then transport's goal might be in the region of 20 to 25 per cent.

This chapter examines the policy measures that are available for controlling CO_2 emissions from personal travel and which would be likely to make a substantial contribution to an overall CO_2 reduction strategy.

Approaches to controlling CO_2 emissions from personal travel

A broad spectrum of policy measures exists for reducing CO_2 emissions from personal travel. In order to make some sense of them, it is useful to categorize them into two general approaches, each containing a range of measures. The first of these categories, Approach A, might be termed 'technical fixes'. It includes measures designed to reduce CO_2 emissions within the existing pattern of travel and modal distribution, by:

- more efficient use of fossil fuels; and
- switching to energy sources that produce less, or no, net CO_2.

The second category, Approach B, involves more radical policy measures, designed to rearrange the present distribution of travel between modes to one that uses less energy. This is achieved by:

- modal shifts towards more efficient transport types; and
- reducing the overall demand for travel.

The two approaches, A and B, are by no means exclusive, but represent two distinct philosophies. Approach A tackles the problem of greenhouse emissions by technological means alone, without addressing the nature and volume of transport demand. Approach B is politically more radical, aiming to modify the volume and the characteristics of travel itself. A successful strategy may need to draw upon policies from both categories.

A continuation of present policies such as that modelled in Scenario 1, 'business as usual', implies no deviation from present trends in travel volume, energy consumption and CO_2 emissions. If current trends are to be broken, there must be some stimulus for this to happen, based on Approaches A and B.

Green consumerism

Since the late 1980s, there has been evidence of a shift in consumer behaviour towards more environmentally sound practices – a phenomenon sometimes termed 'green consumerism'. This is where the public pays a higher price for a product or service because of its environmental qualities. In effect, it reflects the fact that consumers are beginning to attach market values to the environmental attributes of products and services. Former British environment secretary Christopher Patten (1989) has described the effect thus:

> There is an ethical dimension which finds powerful expression in the behaviour of individuals...The green consumer and the green investor are beginning to emerge as a significant force in the market place.

To some extent, green consumerism might be expected to deliver a significant improvement in the environmental impacts of personal travel, as the public begins to 'vote with its feet' for more environment-friendly ways of travelling. Examples might include a greater demand for more economical cars, or an expansion of cycling as a mode of travel.

There is, however, a flaw in green consumerism as an instrument for environmental protection. Consumers attach values only to those environmental assets which they know about and with which they are concerned. Many forms of environmental damage may be overlooked by the consumer, either because information is incomplete or because concern is lacking. Concern for environmental issues does not, unfortunately, always correspond to their ecological importance.

Policy 'levers'

Environmental benevolence on the part of consumers is thus not a sufficiently powerful instrument for change. Instead, policy 'levers' are needed to bring about more substantial changes in behaviour. To reduce CO_2 emissions, as in other areas of environmental concern, there is some debate as to what the stimulus for change should be. There are three broad categories:

■ *Government regulation*, whereby standards relating to environmental protection are set nationally or internationally. Society as a whole, or sectors thereof, are legally bound to observe the prescribed limits. This approach is also termed 'command and control'.

- *Market-based instruments*, in which the state uses market forces by modifying the price of environmentally sensitive activities; the 'polluter pays principle' applies, whereby there is no regulation of environmental damage but instead a cost is assigned according to the severity of the damage.
- *Investment* in less environmentally damaging services and facilities.

The following sections examine the general principles of these three different policy levers, and examine in detail their applicability in reducing CO_2 emissions.

Government regulation

In many cases, environmental goals can be achieved through the application of government regulations. These may be applied to businesses, industries or consumers, and they take the form of a mandatory requirement to follow prescribed procedures. For example, the EC legislated in 1989 that all new cars sold in Europe must be fitted with three-way catalytic converters from 1993. Similarly, the annual check to which cars in Britain are subjected requires that car engines be tuned so as to limit exhaust emissions to a specified level. Such policies have demonstrably succeeded in their stated objectives.

But regulations are not always the best way to solve environmental problems. They tend to demand the same action from all parties regardless of the cost incurred, and may be seen as an inequitable means of achieving environmental targets. For example, larger companies are often better placed to respond to new regulations than smaller ones. Similarly, a catalytic converter imposes a proportionately larger cost penalty on a cheap car than it does on a larger, more expensive model. Regulations can also tend to discourage technological innovation, as they are often set on the basis of what existing technology can deliver, rather than on the standards that competing companies might achieve given the incentive to apply the full weight of their technological expertise.

Market-based instruments

Market-based instruments can often provide a more effective alternative to regulations in achieving environmental objectives. Market forces may be harnessed either to achieve environmental goals or to compensate for environmental impacts, directly or indirectly. Cairncross (1991) cites two examples:

> Individuals may drive their cars to work, rather than take a bus; companies may use chlorofluorocarbons in their commercial refrigerators. In both cases, the costs to society at large, from traffic fumes in one case and from a damaged ozone layer in the other, exceed any private cost to individual or company. That is inefficient. Governments need to step in to align private costs with those to society at large.

Many traditional arguments in favour of a market-based approach to environmental protection were highlighted in 1989 by 'The Pearce Report' (Pearce et al, 1989). This proposed that the natural environment be subjected to traditional market mechanisms, using a cost-benefit approach, with monetary values assigned to environmental assets. Accordingly, the price of all goods and services would – in theory at least – reflect the full environmental costs involved in their provision.

In many cases, market-based measures are already in operation in the form of revenue-raising taxes, and a manipulation of the existing taxation structure could therefore address a number of environmental goals. Transport, and in particular road transport, is already subject to a variety of taxes, which could be adjusted to provide market-based policy levers. Potter (1992a) points out that very few transport taxes were instituted with policy goals in mind:

> Historically, fiscal policies have only been used to raise money for government finances and as part of broad economic and employment intervention. Only very rarely have fiscal measures been used to promote transport policy goals, even though their impacts on transport are substantial.

If existing taxes could be redirected to achieve policy objectives, environmental targets could be addressed without affecting the overall tax burden. In the words of Martin and Michaelis (1992), 'If car users had to pay the full external cost, in a system which recovered the full costs of the congestion and nuisance effects of car use, they would be far more likely to take the train.'

A market-based approach to environmental protection can be found in the principle of 'environmental taxation', which would replace existing fiscal arrangements by taxes that reflect environmental impacts. The overall level of taxation would be the same before and after the reform (ie fiscally neutral). Whitelegg (1991) explains the principle, which is termed 'ecological taxation reform' or ETR:

> Ecological taxation is based on the principle that taxes should fall most heavily on those activities and materials that produce pollution and/or environmental damage. Such taxes would replace taxes on labour and capital, would be phased in over 30 years, and would exert a steering effect on the economy so that environmentally damaging activities (e.g. carrying freight by lorries) would gradually be replaced by environmentally-friendly alternatives (e.g. rail). ETR is not an additional tax: it is a replacement tax. The total taxation burden would remain constant.

Carbon tax

In the context of energy consumption, environmental taxation could take the form of a carbon tax, paid by consumers of fossil fuels in proportion to the amount of carbon they burn. This means that carbon-based fuels would be priced in such a way as to internalize the environmental cost of increased radiative forcing (Barrett, 1991). In September 1991 the European Community proposed a form of environmental tax, in the shape of a hybrid between a carbon tax and an energy tax. Half the revenue would be based on the energy content of fuels, and the other half on the carbon content. The EC carbon tax would involve $10 being added to the cost of a barrel of oil by the end of the century (Gardner, 1991).

Unfortunately, disagreements between member states have delayed the introduction of the tax. Britain, in particular, has made clear that it would rather pursue its own national policy for the taxation of fossil fuels than fall in line with a Community-wide scheme. In his 1993 Budget speech, Chancellor of the Exchequer Norman Lamont said of the carbon tax proposals:

> There may indeed be a case for further coordinated international action on global warming. But I remain unpersuaded of the need for a new European Community tax. Tax policy should continue to be decided here in this House – not in Brussels.

Political difficulties aside, the exact level at which a carbon tax should be set is subject to a number of variables which, when combined, lead to considerable uncertainty. The price of crude oil can fluctuate significantly, even without the addition of a carbon tax. Furthermore, estimates of how the demand for fuel will respond to a given increase in price are far from certain. Small changes in the assumed elasticity of demand can have a profound influence on the necessary level of carbon tax. Barrett (1991) quotes widely varying estimates of the savings that different levels of carbon tax can achieve, and concludes:

> While it is difficult to compare the estimates from one study with those of another, the qualitative story is pretty clear. To lower CO_2 emissions very substantially would require a large carbon tax – larger, certainly, than the taxes already implemented, or for which there exist firm proposals.

Because of the uncertainties involved in the introduction of a carbon tax, there is a strong argument for an 'iterative', target-led approach, whereby the tax is set at a level that is judged to be approximately suitable, then adjusted annually on the basis of the observed response. This principle is expounded by Potter (1992a), who holds the view that '... using the fiscal system to achieve non-fiscal goals is such an uncertain area that it is expedient to introduce measures and adjust them as their effectiveness is determined'.

Tradeable permits
As indicated earlier, different individuals are likely to have a varying ability, or desire, to comply with environmental targets. The principle of tradeable permits allows for these differences to exist, by enabling individuals who exceed the required reduction in environmental damage to 'sell' their excess to individuals who have not managed to do

so. The government might issue 'pollution permits' to individuals, companies or establishments, which can then be bought and sold at an agreed price. For example, companies that do not wish to comply with government pollution targets would be able to 'buy' additional permits from another company. Companies that reduced their pollution output would have spare permits that they could sell to the polluters.[1]

Tradeable permits may be used to introduce an element of flexibility into a market-based programme of environmental protection, and although somewhat regulatory in nature, they work via the market mechanism of altering the supply side of the supply-demand balance. Permits may be used on a variety of scales, from regional to international. In the USA, the Environmental Protection Agency has demonstrated two examples of the use of tradeable permits to curb emissions from the industrial sector. In one of these, oil refiners were allowed to trade credits during a mandatory reduction in the lead content of petrol, with an estimated saving to consumers of $200 million per year (Miller, 1990). Another example of tradeable permits can be found in the Californian car market, where manufacturers will soon be required to sell a certain proportion of 'clean' vehicles every year.

Grants and subsidies

Another form of market-based instrument that can be used to influence environmental impacts is the regulation of prices through government grants. National governments can, in addition to imposing taxes, subsidize activities that will reduce environmental impacts. In transport, this might entail a subsidy of the revenue raised by public transport operators from fares, thereby allowing lower and more attractive prices to be set. Subsidized fares have been widely used in the past, generally for non-environmental reasons. But they can be an expensive means of achieving policy goals, and may be politically contentious. For example, the Greater London Council's 'Fares Fair' policy of the early 1980s, which involved substantial cuts in public transport prices, was the focus of a major conflict between the council and the Conservative government, and ultimately contributed to the abolition of the GLC and the metropolitan councils. More recently, the possibility of a reduction or abolition of the revenue subsidy given to loss-making sectors of British Rail has raised vociferous opposition from a wide range of interests.

Reductions in public transport fares may not be as environmentally beneficial as they appear, as they do not necessarily lead to transfers away from the car to other modes. A major effect is the generation of *additional* trips among people who were already using public transport. On both political and environmental policy grounds, revenue subsidy therefore needs to be used with caution.

Investment

Alongside regulations and market mechanisms, investment can play a significant role in reducing environmental impacts. The term 'investment' is taken here to mean the allocation of government funds in order to facilitate capital spending programmes such as expanded infrastructure, increased capacity and improved services. Investment in public transport generally leads to increased patronage, since transport operators are able to offer an improved service or a greater capacity, or both. In Britain, the government has begun to promote the use of joint public-private sector funding for major transport infrastructure projects, and investment capital for formerly public-sector transport developments has been forthcoming from a number of private companies. Privately funded projects have included the Channel Tunnel rail link and the proposed CrossRail link running east–west beneath London, as well as road and bridge schemes such as the second Forth road bridge.

Summary

In summary, regulations, market mechanisms and investment can all be exploited to curb CO_2 emissions from personal travel. It is impossible to make a generalized choice between regulations and market mechanisms as a solution to environmental problems. Different sectors warrant different solutions, and the choice between regulations and market instruments must be made on a case-by-case basis. For example, the building industry is likely to respond better to energy-saving regulations than to energy taxes, since the value of saved energy is generally small when compared with the initial cost of constructing a house. The transport sector, on the other hand, has a rapid turnover of vehicles, and can respond more quickly to financial incentives. But in general, market-based incentives have greater flexibility than regulations. In a typical market-based arrangement, individuals who are unable or unwilling to comply with standards can instead pay a levy, whilst those who exceed the standard are rewarded.

The three types of policy lever are now assessed in the context of Approaches A and B. Chapter 6 considers Approach A, 'technical fixes', and examines the technological measures available for reducing CO_2 emissions from personal travel. Using the SPACE model a second time, it considers a modified version of 'business as usual', Scenario 2, in which travel patterns are unchanged but technological advances are introduced to reduce CO_2 emissions from individual vehicles. The aim of Chapter 6 is to establish whether, and to what extent, technology alone can provide a means to achieve significant reductions in travel-related CO_2 emissions.

Notes

1. A fuller discussion of this subject can be found in Markandya (1991).

6

Prospects for 'Technical Fixes'

Efforts to improve energy efficiency will have to be strength-
ened if the goal of a significant absolute reduction in CO_2
emissions from energy-use is to be achieved.

Lee Schipper, Lawrence Berkeley Laboratory

The previous chapter showed that technological solutions to transport's
CO_2 emissions could be divided between measures to improve energy
efficiency (fuel economy), and transfers away from fossil fuels to energy
sources which produce less or no CO_2 (alternative fuels). These two
distinct categories can both be exploited to yield an overall reduction in
CO_2 output per kilometre travelled.

This chapter will explore the potential for reducing CO_2 emissions
through fuel economy and alternative fuels, and assess the likely effect of
policies designed to promote these technologies. The SPACE model will
be used to construct a second scenario, similar to Scenario 1, in which
'technical fixes' are exploited to their maximum feasible potential.

Vehicle fuel economy

When considering the possible benefit of fuel economy improvements, it
is useful to concentrate on cars, because:

- Cars are the largest producer of CO_2 within the passenger transport
  sector, responsible for around 90 per cent of emissions (see Chapter 3).
- The potential for improving fuel economy in cars is greater than in
  other modes. Generally, public transport vehicles tend to have a fuel
  economy that is close to the maximum cost-effective level.

The SPACE model, running a 'business as usual' scenario, showed that a
small ongoing improvement in car fuel economy will be insufficient to
curb CO_2 emissions from personal travel, since the growth in traffic
volume will more than cancel out the small gain in efficiency. To begin to

tackle emissions, more significant improvements in fuel economy will be required.

Martin and Michaelis (1992) have estimated that CO_2 emissions from cars could be reduced by at least 20 per cent without forgoing comfort and performance, through the use of improved vehicle design and engine technologies. If smaller, more lightweight cars were introduced, they believe that the saving could rise to over 50 per cent. Estimates of the possible fuel savings that could be achieved through individual technological improvements are given in Table 6.1.

Table 6.1 Technologies available for improving car fuel economy

System	Technology	Fuel saving (%)
Engine design	Precision cooling, reduced engine friction, reduced pumping losses	Up to 6
Power plant	Four valves per cylinder, four-stroke	5 to 15
	Direct injection 2-stroke	Up to 10
	Diesel engine	20 to 30
	Electronic engine management	5 to 20
Vehicle and transmission design	Automated manual	10 to 15
	Continuously variable transmission	10 to 15
	Weight reductions	15 to 20
	Aerodynamic improvements	5 to 10
	Tyre, lubricant and accessory improvements	5 to 10

Source Martin and Michaelis, 1992

Green consumerism, as discussed in the previous chapter, has made little impact so far on the market for vehicles. In the climate of low oil prices that has prevailed in the last 20 years in most countries, consumers have tended to value other attributes, such as comfort, vehicle size, performance, image and so on more highly than the financial and environmental benefits of conserving energy. Figure 6.1 shows that average car fuel economy in Britain, along with most other OECD countries, has hardly changed in the last 20 years, despite significant advances in vehicle technology.

It is likely that the potential of green consumerism for promoting fuel economy could be improved by the introduction of better product labelling, since lack of information is one reason why individual goodwill is an unreliable instrument for achieving policy goals. At present, car dealers are required by law to make available – but not necessarily to display – information on the fuel economy of their products for customers. In practice, environmental aspects such as fuel consumption tend to be downplayed in favour of vehicle size, comfort and performance.

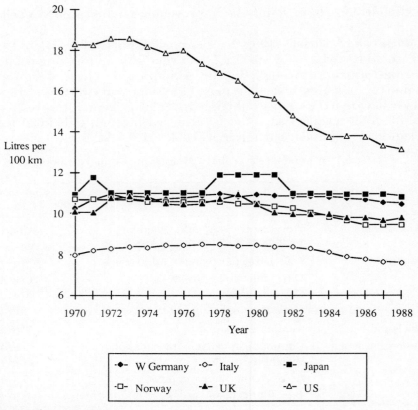

Litres per
100 km

Source Schipper, 1991

Figure 6.1 Average car fuel consumption in selected countries, 1970–88

But poor information is only part of the picture, since consumers may not make 'environment-friendly' transport choices even when equipped with perfect knowledge of the issues involved. Positive incentives are required in order to promote sales of more economical vehicles. A variety of measures are available for promoting fuel economy as an issue for the car buyer, including both regulations and market mechanisms.

Regulations

In the USA, experience over the last two decades suggests that regulations can be effective in stimulating improvements in vehicle fuel economy. In the mid-1970s, following a tripling in world oil prices, the federal government established the concept of Corporate Average Fuel Economy (CAFE), a measure representing the average fuel economy of new vehicles being sold by each car manufacturer. Under the Energy Policy and Conservation Act of 1975, mandatory fuel economy standards were introduced for cars and light trucks (Greene, 1989).

Beginning at 18 miles per US gallon (13.1 litres per 100 km) in 1978, the passenger car standard has progressively increased, and now stands at 27.5 mpg (8.6 litres per 100 km).

Figure 6.1 illustrates the trend in average car fuel economy in the USA since 1970, compared with that of other OECD countries. The average fuel economy of cars entering the market has doubled, whilst that of cars actually in use has improved by almost as much. An extensive analysis of energy use in transport has revealed that average vehicle fuel economy is virtually constant in all OECD countries, except the USA and countries where diesel cars are increasing their market share (Schipper et al, 1991).

Greene (1989) considers the extent to which the CAFE legislation was responsible for the striking improvement in fleet average fuel economy, and whether increases in fuel price may have been a more significant factor in the move towards more economical cars. His approach is to divide car manufacturers into two categories: firstly, those for whom the regulations required an increase in CAFE (mainly domestic car makers); and secondly those who were largely unaffected by the regulations, since their cars were already exceeding the minimum standard set by the government (mainly overseas manufacturers). The two categories are termed 'constrained' and 'unconstrained' respectively. Examining the effect of fuel price on the fuel economy of new cars in either category, Greene's study shows that:

> Constrained-carmaker mpg was twice as sensitive to rising prices as falling prices, suggesting that manufacturers took some actions to counteract the downward pressure of falling prices in order to meet the targets. Unconstrained manufacturers, in contrast, responded slightly more to falling than to rising prices...These results [among others] support the assertion that the CAFE standards were effective in influencing many carmakers to plan for and achieve dramatic increases in new car fuel economy.

Further evidence for the effectiveness of the CAFE standards comes from the observation that average fleet fuel economy in the US has continued to increase even during periods of falling fuel price.

The principal drawback of fuel economy regulations is that they do not necessarily reduce overall petroleum consumption. Even though new cars may be more economical than before, there is no incentive to reduce the overall amount of fuel actually consumed. In fact, consumption may increase as better fuel economy makes motoring cheaper, and encourages greater use of cars (von Hippel and Levi, 1983).[1] Indeed, petroleum consumption by cars in the USA has shown a gradual increase since 1970, despite considerable improvements in the average fuel economy of individual vehicles.

Market-based schemes and taxation

Market-based schemes, on the other hand, could overcome the price effects that tend to increase travel demand. For example, the use of *fuel taxes* to promote fuel economy would mean that the amount spent on fuel in terms of pence per kilometre could remain the same, with the increased fuel price stimulating improved fuel economy. As cars became more economical, the fuel price penalty per vehicle kilometre would diminish.

Rearranging the transport taxation system to include environmental taxation principles could encourage the use of more economical vehicles. In Britain, Vehicle Excise Duty (VED) is charged at a flat rate of £125 per year for private cars. By charging VED at a rate that is directly related to the fuel economy of the car, it would be possible to provide consumers with an incentive to purchase more economical vehicles, with no effect on overall revenue raised. Other motoring taxes could similarly be varied according to fuel economy. In the words of the *London Evening Standard* (1991),

> Vehicle tax is an objectionable flat-rate tax, levied on Porsches and Ford Cortinas alike. Slapping an increase on petrol costs would automatically penalize the gas-guzzlers which do most damage to the environment. A petrol tax is, therefore, a green tax: it would score brownie-points with the environmentalists while not diminishing the revenue which the Exchequer would receive.

Some researchers, particularly economists in North America, have taken a dim view of the improvement in average fuel economy that could be brought about through market measures. The following is a typical example, by Cambridge Energy Research Associates:

> For a car driven 12,000 miles per year, a hypothetical mileage improvement from 30 to 45 miles per gallon would save 150 gallons of gasoline annually. The cost of achieving this saving through redesign can only be estimated, but, on plausible asumptions, is likely to be on the order of $0.70 per gallon saved. At a fuel cost of $1 per gallon, the net saving to the driver is only about $0.30 per gallon saved, or only about $50 a year.
>
> Cragg (1992)

However, this line of reasoning is at best simple minded, and at worst downright misleading. The notion that switching to a more economical car will actually cost the owner money bears little resemblance to the real world, where, as everyone knows, cars that are economical on fuel also tend to be smaller and cheaper to run.

In short, the critics are assuming that fuel economy improvements will be brought about by redesigning cars whose speed and performance are the same as those currently in use. In other words, they are precluding the possibility that some drivers might be prepared to give up a certain amount of horsepower in order to reduce their fuel consumption. This implicit assumption places an artificial and misleading limit on the level of fuel economy that might be considered achievable in the future.

As indicated earlier, fuel economy improvements alone do not necessarily yield reductions in fuel consumption. Motorists may tend to use their cars more as a result of the reduced fuel costs associated with more economical vehicles. To remove this feedback effect, VED and other 'flat rate' taxes could be eliminated and the revenue collected instead through increased fuel taxes. The use of fuel taxes as a means of raising revenue encourages consumers not only to purchase more economical vehicles but also to reduce their overall travel volume.

Moreover, changes to the existing taxation structure could be made 'revenue neutral', so that the overall tax yield is unchanged. This can be a useful means of improving the likelihood of public acceptability.

Since 1978 the CAFE standards in the USA have been complemented by the 'Gas Guzzler Tax', a one-off levy charged on cars which consume 10.5 litres per 100 km or more. The charge varies according to the margin by which the vehicle falls short of the standard. In one sense, the tax has been a victim of its own success: manufacturers of large, fuel-thirsty cars have taken steps to move inside the 10.5 litres per 100 km threshold, and revenue generated by the tax is now relatively small. The principal drawback of the Gas Guzzler Tax is that it provides no incentive whatsoever for manufacturers to improve the fuel economy of a product once it is safely below the prescribed cut-off level (European Federation for Transport and Environment, 1992).

Another type of market-based measure that uses restructured taxation to encourage fuel economy can be found in the form of 'feebate' programmes. These are revenue neutral schemes whereby consumers who purchase economical vehicles are given rebates, whilst those who do not are charged fees. As before, no mandatory standards are imposed, but consumers are instead required to pay the price of 'undesirable' practices.

A feebate scheme has been proposed in California, aimed at stimulating sales of 'cleaner' cars. Termed 'DRIVE+' (Demand-Based Reductions in Vehicle Emissions Plus Improvements in Fuel Economy), the programme addresses all forms of pollution from vehicle exhausts, not only CO_2. DRIVE+ would impose sales tax surcharges on purchasers of large or polluting vehicles, and the proceeds would be used to fund rebates for purchasers of less polluting vehicles. The entire programme would be revenue-neutral and require no support from government (Levenson and Gordon, 1990).

A similar arrangement has already been established in Ontario, Canada. The original plan proposed a system of sales tax that was related to the fuel economy of the vehicle, with the least economical category attracting a tax of $7,000. However, pressure from the motor industry resulted in a modified version of the scheme being implemented. The tax rates for mid-range vehicles were made less severe, although the maximum fee remained unchanged. The revised scheme offered a rebate of $100 to purchasers of the most economical vehicles. Table 6.2 gives details of the programme.

Table 6.2 Taxation of passenger vehicles under the Ontario programme

Highway fuel economy (litres per 100 km)	*Tax/(Rebate)*
Less than 6.0	($100)
6.0–8.9	$75
9.0–9.4	$250
9.5–12.0	$1,200
12.1–15.0	$2,400
15.1–18.0	$4,400
More than 18.0	$7,000

Source Laughren, 1991

A scheme similar to the Ontario programme could be established in other countries, based on the official test figures for new-car fuel economy. Potter (1992a) suggests that car purchase tax, VED and VAT might all be related to vehicle fuel economy, using a system of rebates and surcharges. This could be done using either a threshold 'minimum' fuel economy or, more usefully, a banded structure similar to that established in Ontario.

While there is a wide range of possible fiscal measures to stimulate fuel economy in the car sector, there has been only limited experience of such policies in action. Japanese trials with vehicle pricing support the view that market mechanisms can stimulate rapid changes in the vehicle fleet. In 1976, *kei* (lightweight) cars were made exempt from on-street parking charges in Japan. (*Kei* cars have a maximum engine capacity of 550 cc and a maximum length of 3.2 metres.) By 1988 sales of *kei* cars had soared, with 1.8 million vehicles in this category being sold per year. Unfortunately, the market for these vehicles began to decline in 1989 following the abolition of the parking exemption (*The Economist*, 1990).

In Italy, a number of fuel economy-related tax measures have been adopted. Purchase tax has traditionally been higher on cars whose engine capacity exceeds 2000 cc (2500 cc for diesels). In addition, the property tax that is payable every year on cars by their owners is related to a banded structure representing 'fiscal horsepower'. These market-based policies, similar in effect to 'carbon taxes', have had a significant effect upon the fuel economy of private cars. Car fuel economy in Italy since 1970 has been better than in any other European country. In 1988, it was 23 per cent higher than in Britain (Hughes, 1991).

In Greece, tax breaks favouring smaller-engined cars were introduced in 1990 as a means of reducing the chronic air pollution problems faced by cities such as Athens. A significantly lower rate of sales tax was made available on cars below 2.0 litres in engine capacity, as well as five years' exemption from road tax. To qualify for the discount, the car purchaser was also required to dispose of an older vehicle, whose output of polluting gases was likely to be greater. Around one thousand old cars were scrapped during the first month of operation (Cragg, 1992).

Another example of taxation measures used to promote fuel economy can be found in the Isle of Man, a part of the British Isles which is not subject to the legislation of the British government. The annual licence fee for cars is related to engine capacity, according to six categories. These are detailed in Table 6.3. Although the annual fees are not large when compared with the annual cost of owning and operating a car, the Isle of Man scheme nevertheless provides an example of a system that could be operated in virtually any country.

Table 6.3 Annual licence fees for cars in the Isle of Man

Engine capacity (litres)		Licence fee
More than	*Not more than*	
	1.0	£38
1.0	1.2	£45
1.2	1.8	£55
1.8	2.5	£70
2.5	3.0	£105
3.0		£115

Source Isle of Man government

Finally, there is also scope for tradeable permits to be used as a policy lever to promote sales of more economical cars. These would apply pressure mainly to the supply side of the car market, by giving manufacturers 'credits' for selling economical cars. Manufacturers wishing to continue selling 'gas guzzlers' would first have to buy the necessary credits from another company. In 1992 the British government published a paper approving in principle the idea of using tradeable credits to control CO_2 emissions from transport (Department of Transport, 1992b).

In summary, there is a broad spectrum of market mechanisms that might be used to promote vehicle fuel economy. In most countries, the present system taxes car ownership relatively heavily, while taking little account of car usage and its environmental impacts. The measures described above offer a means of incorporating the environmental cost of CO_2 emissions into the taxation system.

The improvement in average fuel economy that has taken place in the USA since 1975 is a testimony to the effectiveness of regulations. Other successes for regulations include the phasing-out of CFCs in accordance with the Montreal Protocol and the adoption of EC-wide vehicle emissions standards in 1989. But in other areas market mechanisms may be preferable, because:

- CAFE-type schemes do not encourage manufacturers to go beyond the level of fuel economy required by the regulations.
- In the present political climate, market-based solutions are more likely to be acceptable than additional regulations. Market-based instruments can cost considerably less to central government, particularly if a revenue-neutral scheme is adopted.

■ Fuel economy regulations can be successful in improving specific energy consumption (SEC), but may not necessarily contribute to reduced energy consumption overall. As shown above, reduced fuel costs can lead to increased car usage. A market-based scheme, on the other hand, can be used to charge motorists for their *overall* fuel consumption, rather than just the specific fuel consumption of their vehicle.

Although the CAFE standards introduced in the USA appear to have made a major contribution to improving fuel economy, the extent to which these improvements are attributable to the CAFE standards will probably never be understood with confidence. However, the main criticism of the American approach is that market measures, in the form of fuel taxes, could have achieved the same result at a fraction of the cost. In the words of Leone and Parkinson (1990):

> A market-based approach to energy conservation is superior in effectiveness to regulation of the corporate average fuel economy (CAFE) of new vehicles. Market incentives to encourage energy conservation cost society far less than command and control regulations like CAFE.

Other measures

As well as the design of the vehicle, *speed* can have a major influence on fuel consumption, as illustrated in Figure 3.12. At the upper end of the speed range, fuel consumption increases broadly with the square of speed. In Britain the national speed limit is 70 mph (113 kmh) on motorways and dual carriageways, with lower limits for other types of road. However, there is evidence of widespread flouting of this regulation: a Department of Transport survey in 1992 indicated that 56 per cent of cars at free-flowing locations on motorways were exceeding the legal limit of 70 mph (Department of Transport, 1993b).

A significant volume of fuel could be saved simply by enforcing the national speed limit. Full enforcement of Britain's 70 mph speed limit could result in the saving of around 500,000 tonnes of petroleum by cars alone, corresponding to the elimination of 1.5 million tonnes of CO_2 each year.[2] The elimination of excessive speeds can also reduce congestion at bottlenecks, thereby reducing the amount of fuel lost as a result of inefficient driving in such conditions. The British government announced in 1993 that it is to experiment with the use of 50 mph (80 kmh) speed limits on the highly congested M25 orbital motorway around London, to replace the present limit of 70 mph. Kenneth Carlisle, roads minister at the time, justified the introduction of a lower speed limit as follows:

> Our research and experience from overseas suggests that this action stabilizes the traffic flow and helps prevent stop-start conditions from

developing. In this way, the risk of major hold-ups can be reduced, and so average speeds throughout the peak may well increase.

Measures to bring about the better enforcement of speed limits could be deployed almost immediately, but would incur costs in terms of additional policing time and surveillance technology. Roadside equipment is available for monitoring vehicles electronically, and photographing those that exceed the limit. Such devices are now permitted by law, but are expensive. A single camera costs around £10,000, but there is evidence that 'dummy' cameras can also be effective in deterring drivers from speeding. A number of police authorities have adopted the technique of moving a single camera between a number of sites, with dummy flashes used when the camera is not in place. This gives speeding motorists the impression that they have been photographed, even if they have not.

Alternative fuels

The second area in which technological changes may be used to reduce CO_2 emissions from individual vehicles is alternative fuels. The term is used here to cover not only different types of fuel, but also novel engine types. The following sections examine in detail the options that are currently available for alternative forms of propulsion, and policies that might be used to promote these technologies. It is critical to distinguish between alternative fuels which are of use in curbing CO_2 emissions, and those which are not. The following assessment takes into account not only energy consumption at the point of use, but also the processes by which the fuels are produced and supplied to the point of use.

Alternative fuels have been advocated world wide for various reasons, and it is possible to identify three distinct motives:

- Conserving oil supplies, or reducing dependence on imported oil.
- Reducing air pollution in areas of high traffic density.
- Reducing emissions of greenhouse gases.

Traditionally, the first two of these have been the main incentive for developing alternative fuels. For example, in 1975, following a world oil crisis, Brazil embarked upon a programme aimed at replacing most of its imported gasoline with ethyl alcohol, distilled from domestically grown sugar cane (Weiss, 1990). Similarly, Canada and South Africa have constructed plant for manufacturing motor fuels from coal and tar sands, while New Zealand has replaced almost half of its petrol consumption with natural gas (Sperling and DeLuchi, 1989). Southern California, gripped by a worsening air pollution problem, has adopted regulatory legislation that will require 40 per cent of new cars to run on 'clean' fuels by 2000, and 100 per cent by 2008 (South Coast Air Quality Management District, 1989).

Many of the technologies developed as part of alternative fuels programmes are relevant from the perspective of reducing CO_2 emissions. However, it is vital that a proper evaluation be carried out, so that fuels with a lower CO_2 output may be separated from those that will give no reduction in emissions. In principle, alternative fuels can be applied to all motorized forms of transport. For example, rail transport has progressed from coal to diesel as its dominant fuel, and more recently electric traction has taken over a large portion of operations. Aircraft, too, are capable of running on fuels other than petroleum.

Electric vehicles

Electric vehicles have been in limited operation for many years in Britain, in the form of electric milk delivery vehicles. They are also used extensively for applications such as golf carts, forklift trucks and indoor people-movers. They store electric charge in batteries carried on board the vehicle which are charged from the mains supply when the vehicle is not in use.

The principal disadvantages of electric traction are the limited range of the vehicle between charges, and the mediocre performance when compared with conventional heat-engined vehicles – which is partly a result of the large mass of batteries that needs to be carried around. Another problem is battery lifetime: the owner of an electric car would need to replace the batteries typically every two or three years.

The benefits of electric traction are most noticeable in an urban environment, where frequent stops are made. In this type of driving, internal combustion engines are at their least efficient. Other benefits of using electric vehicles in cities include reduced air pollution and noise. Furthermore, the limited range of electric vehicles presents less of a problem in low-speed, short-distance applications.

Recent air quality legislation in the USA has prompted renewed interest in electric vehicles. The 1990 Clean Air Act Amendments (CAAA) include a programme that requires the car fleets in a number of heavily polluted areas to contain a percentage of low-emission vehicles. Electric cars are classified as 'zero-emissions vehicles'. In addition, the State of California has ruled that car manufacturers must, from 1994 onwards, comply with increasingly strict pollution specifications that will involve the sale of a certain percentage of electric vehicles (Chang et al, 1991).

Probably the most advanced purpose-built electric car to be produced by a major manufacturer is the Impact, developed by General Motors and launched in January 1990. But Impact had a large number of predecessors which got no further than prototypes, including the Leyland Crompton electric car which was designed in 1972 in Britain. Like the Leyland Crompton, Impact is powered by conventional lead-acid batteries, but it minimizes the traditional handicap of short range by employing lightweight materials and a highly aerodynamic shape.[3]

An increasing number of car manufacturers have joined the race to develop electric cars suitable for the USA market. Ford, Peugeot,

Volkswagen, BMW, Renault, Lada, Daihatsu and Nissan have produced prototypes which they claim can provide levels of range and performance comparable with those of conventional internal combustion (IC) engined vehicles.

BMW's prototype, the E1, uses a sodium-sulphur battery, which the company has spent around £330 million developing (Cragg, 1992). Italian car giant Fiat has launched two electric versions of its Cinquecento hatchback, anticipating restrictions on the use of internal combustion engines in cities. One is fitted with a lead-gel battery, and the other with nickel-cadmium. They cost over twice as much as the conventional petrol-engined version, and have a range of 70 km and a top speed of 80 kmh. Fiat believes that sales will begin to pick up in 1996, when a viable market is expected to become established for electric cars in Italy.

The PSA Group in France plans to launch a range of electric Citroens and Peugeots during 1994. Like Fiat, PSA believes that worsening air quality in European towns will sooner or later lead to restrictions on the use of internal combustion engines. Both Fiat and PSA predict sales of around 200,000 electric vehicles a year in Europe by the year 2000 (Griffiths, 1992).

Ford's Ecostar electric vehicle, based on the Escort van, is powered by a sodium-sulphur battery, and is claimed to have a range of 175 km, and a top speed of 110 kmh. An initial production of 105 vehicles has been undertaken in the USA. Other protoypes developed to date include an Opel Kadett running on nickel-cadmium batteries, a Chrysler TEVan (nickel-iron), a Volkswagen Jetta (sodium-sulphur) and various Toyotas powered by zinc-bromide batteries.

Both sodium-sulphur and nickel-cadmium batteries have better performance characteristics than a conventional lead-acid battery, but cost around six times more at present. A major drawback of the sodium-sulphur battery is that it needs to be kept at a temperature of 300–350°C. In addition, the contents must be heavily protected from accidents, since sodium is a highly reactive metal.

It is not only private cars and trains that can be run on electricity. Four electric buses entered service in Oxford in 1993, in a joint initiative between Southern Electric and City of Oxford Motor Services. The vehicles are claimed to have a range of 80 km when fully charged.

Research into electric vehicle technology received a major boost in 1991 with the establishment of the United States Advanced Battery Consortium, which brings together the research efforts of the 'big three' US car manufacturers – Chrysler, Ford and General Motors. The consortium has a budget of around $2 billion. In 1993, Saft America was awarded a $35 million contract by the USABC, together with the US Department of Energy and the Electric Power Research Institute, to develop the use of two advanced battery types – lithium aluminium iron disulphide ($LiAl\text{-}FeS_2$) and nickel metal hydride (Ni-MH). According to Saft, Ni-MH could provide up to four times the range of conventional lead-acid battery cars. In the longer term, LiAl-FeS$_2$ batteries, operating at 450°C, could compete with lead-acid while offering the advantages of lower unit weight (Saft, 1993).

The European Commission's THERMIE research programme for developing energy-efficiency technologies is likely to yield further progress in the development of electric vehicles. Two projects involving European cities – JUPITER and ENTRANCE – have secured THERMIE funding to demonstrate 'the rational use of energy in transport', including the use of electric power. Meanwhile the French car manufacturer Peugeot is cooordinating the development of 'public' electric car schemes in a number of cities, including La Rochelle and Tours in France, and Coventry in England, using THERMIE funds. A new research programme known as LIFE has recently been announced, under which a number of European cities are hoping to undertake trials with electric vehicles (*Local Transport Today*, 1993).

Hybrids

The nature of the USA legislation, which requires low-emission vehicles to be used in urban areas but places no restriction on the type of vehicle used elsewhere, means that hybrid cars are likely to find a market niche. Hybrids are typically equipped with both an electric motor and an IC engine: the electric motor is used for driving in town, while the IC engine is used for longer-distance trips. The batteries may be charged either from the mains or from the car's IC engine. Already there are a number of prototype hybrid cars. These include an adapted Volkswagen Golf and an Audi 100 with a 2 litre engine and a sodium-sulphur battery. This can travel 500 km on a tankful of petrol, and a further 80 km using the battery (Cragg, 1992). Another hybrid prototype, the LA301, has been developed in Britain by Clean Air Transport in the form of a two-door hatchback, running on a 650 cc petrol engine and an electric motor (Schoon, 1991).

Hybrid engines have also been tested in buses, as a means of reducing air pollution in the centre of busy, pedestrianized town centres. British operator Transit Holdings, which operates buses in Oxford and Exeter, has recently acquired two midibuses powered by a petrol engine and an electric motor, which are also equipped with regenerative braking. The petrol engine powers a generator, which charges the batteries; and in sensitive areas the petrol engine can be switched off altogether (Millar, 1993).

Electricity generation

As explained in Chapter 3, electricity should be regarded not as a primary fuel, but as a medium for storing energy derived from some other source. The means for generating electricity are varied, and include fossil-fired power stations, nuclear stations, and renewable sources of energy (wind, wave, solar and hydroelectric). The level of CO$_2$ emissions is crucially dependent upon the source of primary energy.

Vehicles powered by non-fossil electricity (nuclear and renewable) offer a complete elimination of greenhouse emissions,[4] whilst electricity from natural gas also offers a benefit as a result of its low carbon content. In terms of CO_2 production, electric vehicles using the present USA power supply offer little or no advantage over IC-engined vehicles, and will become even less attractive if further improvements in the fuel economy of IC-engined cars take place. Running on electricity from coal plants, electric cars would actually worsen greenhouse emissions.

The same analysis can be performed for Britain as for the USA, where the mix of fuels in the electricity supply industry is similar. A report by the Society of Motor Manufacturers and Traders (1990) presents the following analysis:

> Although the process of electric power generation is somewhat more efficient than an internal-combustion engine it is still only of the order of 35 per cent. However, the *overall* (power station fuel to road wheels) system for operation of an electric vehicle includes further losses. The electricity distribution system to the consumer is about 90 per cent efficient and assuming a conventional battery charger system, the charger is about 80 per cent efficient and the battery 85 per cent. Finally, the electric vehicle engine (motor) and transmission are somewhere between 60 and 90 per cent efficient. Taking an average for this of 75 per cent, then the overall energy efficiency of the system is around 16 per cent. Use of regenerative braking can give a 10 per cent improvement in the efficiency of the motor/transmission system, giving an overall figure of 17 to 18 per cent. Compared to quoted *overall* (crude oil to road wheels) energy efficiencies of 18 to 20 per cent for petrol vehicles, and a further two per cent for diesel, this strongly suggests that *the use of electric vehicles would not reduce CO_2 emissions where the electricity was produced by fossil fuel combustion* [author's italics].

Electric vehicles can therefore offer reduced greenhouse emissions, but only if the electricity is provided primarily by non-fossil sources or natural gas.[5] Given the present-day mix of fuels used to generate electricity, electric vehicles are not capable of providing reductions in greenhouse emissions. This is especially true if conventional, IC-engined vehicles are to become more economical than they are at present. To have a viable role in reducing CO_2 emissions, electric vehicles will require a substantial shift away from carbon-based fuels in the electricity supply industry. The UK would need to expand its nuclear or renewable power capacity on an enormous scale in order for electric vehicles to have a role in reducing CO_2 emissions. Given the serious questions – both economic and environmental – that now surround the international nuclear industry, the only viable future for electric vehicles as a solution to CO_2 emissions would appear to lie in a massive expansion of renewable energy capacity. It is possible that electricity generated from solar energy could soon be cheap enough to be used as one source of energy for vehicles. The cost of electricity generated from photovoltaic cells has fallen from \$14 to \$2 per watt since 1980 (Cragg, 1992).

Hydrogen and fuel cells

Another 'clean' fuel for road vehicles is hydrogen, which can be either burned in an internal combustion engine or combined with oxygen in a fuel cell. In both cases, the principal waste product is water vapour. Emissions of other pollutants, with the exception of nitrogen oxides (NOx) in the case of IC engines, are negligible. However, like electricity, hydrogen should be regarded not as a primary fuel, but rather a means of storing and transporting energy.

The cheapest source of hydrogen at present is natural gas – though this source offers no benefit, in terms of reducing CO_2 emissions, compared with using natural gas as a fuel directly. A more useful route, from a greenhouse perspective, is the hydrolysis of water into oxygen and hydrogen according to the equation

$$2 \, H_2O \rightarrow 2 \, H_2 + O_2$$

where both products are in the gaseous state. The most promising means of achieving this is the electrolysis of water, which is done by passing an electric current through a conducting solution. As with battery charging, the electricity used in this process can be derived from a variety of sources including fossil, nuclear and renewable energy. But the energy efficiency of electrolysis is typically low, between 5 and 25 per cent (Sperling and DeLuchi, 1989). Alternatively, hydrogen can be manufactured directly from fossil fuel feedstocks, though this would lead to a net increase in CO_2 emissions compared with conventional fossil fuel combustion.

Prototype hydrogen-burning vehicles have been developed by German car manufacturers BMW and Daimler-Benz, though these have been conversions of existing production models rather than purpose-built hydrogen-powered vehicles. The BMW prototype runs on liquid hydrogen, stored at −253°C, and is claimed to have a range of 300 km.

Hydrogen can also be stored as a compressed gas, as a hydrogen-toluene compound, or in the form of *metal hydride*. The last of these is a compound that releases gaseous hydrogen when heated, and which can be 'recharged' with hydrogen when depleted. The advantage of metal hydrides is that they are inert and non-hazardous, in contrast to the highly explosive hydrogen gas. A metal hydride tank would need to be replenished approximately every 500 km.

Hydrogen gas has a calorific value of 120 megajoules per kilogram, compared with just 48 MJ/kg for petrol. But it requires a large volume for storage, and most researchers favour cryogenic liquid storage instead. Even a liquid hydrogen tank requires considerably more storage space than a conventional petrol tank. There are other technical problems with hydrogen-burning engines, such as emissions and lubrication, which lead Martin and Michaelis (1992) to doubt whether hydrogen is feasible as a transport fuel other than in the long term.

A more efficient use of hydrogen fuel can be made using fuel cells, which combine hydrogen and oxygen electrochemically without undergoing a thermodynamic cycle. The efficiency of the process is thus approximately twice as great, though in practical use it would probably drop to between 40 and 45 per cent. The most promising design for transport purposes is the solid polymer fuel cell, which has a high power density and a low working temperature. Other types include the 'molten carbonate' and 'solid oxide' fuel cells, both of which require higher temperatures and are therefore less suitable for automotive applications. Costs of fuel cells are some ten times higher than those of conventional engines (ibid).

The first fuel cell-powered car was developed in the 1960s by Shell. In 1991, automotive fuel cell technology was significantly advanced by the invention of a new device called the Lasercell, which could be run in reverse in order to generate hydrogen fuel from water. In practical terms, the cell can be connected to the electricity supply when the car is not in use, and the metal hydride tank replenished (Webb, 1991).

In Vancouver, Canada, a fuel cell-powered bus was launched in 1993, as part of a joint venture between manufacturer Ballard Power Systems and the province of British Columbia. The fuel cell is powered not by hydrogen but by methanol, though the operating principle is the same in either case. Air quality officials in Southern California have taken a keen interest in the project, and are considering the use of fuel cells as an alternative to diesel in commuter railways.

As with electric vehicles, the level of CO_2 emissions from hydrogen-powered vehicles depends critically on the source of the fuel. Hydrogen can be produced electrolytically using nuclear or renewable energy, with negligible emissions of greenhouse gases. Fossil fuels may be ruled out as a source of electricity, because hydrogen can be produced *directly* from fossil fuels with a greater efficiency. Sperling and DeLuchi (1989) indicate that:

> Fossil fuels would not be used as the source of electric power because it would be cheaper and more efficient and would generate less carbon dioxide to make the hydrogen directly from fossil fuels. Hence nonfossil feedstocks, such as solar, geothermal, wind, hydro, and nuclear energy would be used to generate electricity for the electrolysis process.

Emissions of CO_2 from vehicles using any form of fossil-based hydrogen are considerably greater than those from conventional petrol or diesel engines. If, however, hydrogen is produced electrolytically using non-fossil electricity, it is possible in principle to eliminate CO_2 emissions completely.[6]

In summary, hydrogen gas has a considerable potential for reducing CO_2 emissions at a competitive cost, but this depends critically on the expansion of renewable energy capacity. Fuel cells, if they become commercially viable, will offer better energy efficiency than IC hydrogen engines. However, fuel storage can present problems, particularly in smaller vehicles.

Alcohols and biofuels

Alcohol fuels have been put forward as a solution to a number of transport problems, particularly those related to air quality and oil supply security. In practical terms, alcohol fuels for transport comprise methanol (CH_3OH) and ethanol (CH_3CH_2OH). Both fuels can be used in conventional internal combustion engines, though modifications are required. In practice the properties of alcohol fuels are best exploited using purpose-built engines.

Alcohol fuels fall into the category of either *synthetic* fuels (synfuels) or *biomass* fuels (biofuels). Synfuels are typically manufactured using coal or oil shale as feedstocks. (As well as synthetic alcohols there are synthetic natural gas and synthetic petroleum.) The production process is energy intensive, and is usually justified on the grounds of oil supply shortages. Sperling and DeLuchi (1989) have outlined the disadvantages of synfuels compared with conventional motor fuels. One observation in particular is of relevance from a greenhouse perspective:

> Possibly the most serious environmental impact is the large amount of carbon dioxide emitted by the use of coal and oil shale-based fuels. [These] produce about twice as much carbon dioxide as petroleum. If the scientific consensus on the seriousness of the greenhouse effect is translated into policy, coal and oil shale-derived fuels could be eliminated by this criterion alone.

Alcohols produced from coal and oil feedstocks can therefore be ruled out of any strategy for reducing greenhouse gas emissions. Methanol can alternatively be manufactured from natural gas, but the reduction in CO_2 emissions compared with conventional petroleum fuel is very small. Sperling and DeLuchi (op cit) estimate that methanol derived from natural gas could reduce emissions by just 3 per cent relative to an equivalent petrol-engined vehicle. Such a small benefit in terms of CO_2 emissions could not justify a switch to this fuel on the grounds of curbing global warming. This is especially true while vehicle fuel economy remains at the current level, since improvements in vehicle fuel economy can offer greater reductions in CO_2 emissions at a lower cost.

The use of methanol is expected to increase in the USA (and in particular California) because of air quality concerns. It is likely to be used in a mixture containing 15 per cent gasoline, which helps the fuel to vaporize in cold weather. Natural gas is favoured as a feedstock for the fuel because of its abundance at low cost. However, emissions of CO_2 would be lower if natural gas were used directly as a motor fuel rather than being first converted to methanol. In short, alcohols based on fossil fuel feedstocks can be discarded as options for reducing emissions of CO_2.

The second, and more promising, option for the manufacture of alcohols is biomass, or plant matter. The potential for reducing CO_2

emissions by switching to biofuels is considerable. The key consideration is that although CO_2 is produced when plant matter is burned, an equivalent volume of CO_2 can be recovered from the atmosphere by replanting the harvested crop, thereby absorbing this gas from the air through photosynthesis.

Two broad categories of feedstock are available for the production of alcohol biofuels. Firstly, crops and food wastes containing large amounts of starch and sugar, such as sugar cane and maize, may be fermented to produce ethanol. Secondly, a form of biomass that is cheaper and more abundant is *lignocellulosic* material, derived mainly from tree products. Cellulose can be either thermochemically processed to make methanol or hydrolyzed to produce ethanol.

Studies in Europe have indicated that bioalcohols are likely to be most useful if blended with conventional gasoline, rather than as a pure fuel as used in Brazil. A number of safety problems are associated with methanol, which is highly toxic and corrosive. It also has about half the energy density of gasoline, which means that a larger mass of it needs to be carried aboard the vehicle.

In the early 1980s, Sweden began experimenting with ethanol as a fuel for buses, exploiting a plentiful supply of waste products from the wood industry as a feedstock for the fuel. Following initial trials, 32 ethanol-powered buses, converted from conventional diesel vehicles, were introduced in Stockholm in 1990 – the world's largest trial of ethanol-powered buses (Smith, 1993). Following the success of the scheme, bus company SL has ordered a further 20 vehicles. By securing a source of energy based on renewable biomass, the Swedish project has demonstrated how measures to improve air quality and those aimed at reducing CO_2 emissions can be complementary.

Another form of plant-based motor fuel besides alcohol is 'biodiesel' – essentially vegetable oil derived from a variety of energy crops. These include palm, coconut, sunflower, rape and soya, and are converted to fuel not by distillation but by simply crushing the plant. Some of these oil crops are already being cultivated successfully in Northern Europe. Most notable among them is oilseed rape, with its distinctive bright yellow colour in early summer. Already rapeseed oil, in the form of rape methyl ester (RME), is providing fuel for a number of bus companies in Italy and Switzerland, as well as in a trial scheme involving three buses in Reading, England. Currently Austria uses a fifth of its rapeseed production for transport fuel, and it is estimated that European production of RME could soon reach 600,000 tonnes a year (McDiarmid, 1992).

Soya oil has a calorific value slightly below that of diesel, but its boiling point is considerably higher and it solidifies at a higher temperature, which implies difficulties with frozen fuel in cold weather. Furthermore, its viscosity is some 10 times greater than that of diesel. These problems typify those of other vegetable oils (Knight and Cooke, 1985). A study of rapeseed oil by Porsche has also revealed some drawbacks in terms of toxic emissions (Cragg, 1992).

But the technical difficulties are not insuperable, and natural oils can provide a diesel substitute that produces significantly less CO_2 than conventional petroleum fuels. In principle, a biofuel plantation which grows as rapidly as the vegetation is removed can have almost zero net emissions of CO_2. In practice, however, a significant amount of CO_2 is produced during the manufacturing process. A study carried out by the German Federal Environment Office (UBA) in 1993 calculated that the CO_2 saving associated with rapeseed oil compared with diesel would be 65 per cent, rather than the theoretical maximum of 100 per cent. Significantly, the UBA study estimated that if other greenhouse gases besides CO_2 were included in the analysis, the advantage of rapeseed over diesel would fall to just 35 per cent (Meyer, 1993).

Surplus agricultural land could be made available for cultivating both oil crops and plant-based alcohols. Presently, throughout the European Community, farmland is becoming redundant as a result of a community-wide rationalization of agricultural output, and there remains a wheat production surplus of around 20 per cent (Martin and Michaelis, 1992). Subsidies are paid to some farmers who leave parts of their land unused, under the 'set-aside' scheme. Large areas of land such as these would be suitable for growing energy crops.

For example, experiments in France have shown a combination of sugar beet and wheat to be a useful source of bioethanol. To switch to a blend of 7 per cent ethanol and 93 per cent gasoline would require 5,100 square kilometres of agricultural land, it was estimated. Meanwhile Sweden has calculated that 50 per cent of its wheat surplus could be eliminated by switching to a 6 per cent alcohol/gasoline blend, and the British government has estimated that there is sufficient excess wheat and barley to substitute for 5 per cent of the country's gasoline consumption (Cragg, 1992).

In addition to the set-aside land, upland areas that are presently planted with coniferous trees might be replanted with fast-growing energy crops. Although there would not be a sufficient supply of natural oils to displace completely the current demand for petroleum, it is likely that several million tonnes of CO_2 emissions could be avoided in Europe alone by switching to plant-based fuels. However, the economics of biofuels are not favourable at present, not least because of the low price of crude oil. A study by the Department of Energy has indicated that oil prices would need to rise to more than $55 a barrel in order for bioethanol to compete with gasoline (ibid). Even then, production would be limited by the availability of suitable land.

In summary, to be useful as part of a greenhouse strategy, ethanol and methanol would need to be derived from biomass. This would be likely to consist of a combination of organic wastes and dedicated fuel crops. The principal limitation on the use of alcohols is the area of land available for growing energy crops. Natural oils, too, offer significant savings in CO_2 emissions. However, in the present climate of cheap petroleum, there are serious obstacles in the way of a large-scale switch to plant-based fuels. But in the longer term, the cost equation is expected to shift in favour of the new fuels as petroleum becomes more scarce.

Natural gas

Natural gas consists mainly of methane, and is found principally in gas wells such as those in the North Sea. It contains less carbon than petrol, for an equivalent energy content, and emissions of CO_2 are approximately 35 per cent less.

Natural gas is used to propel approximately 0.25 per cent of vehicles in Europe, most of which are conversions of petrol-engined vehicles sold in Italy (Dallemagne, 1990). Canada and New Zealand also have sizeable fleets of gas-powered vehicles. World wide, around one million vehicles operate on natural gas, in the form of either compressed natural gas (CNG) or liquefied natural gas (LNG). Of the two, CNG is generally the more popular option, as the production costs are lower than those for LNG. It is burned in a conventional IC engine, and most petrol engines can be adapted to run on CNG.

British Gas has converted around 100 of its vans in Britain to run on either natural gas or petrol. A smaller number of vehicles will be 'dual fuel' conversions, operating on a mixture of natural gas and diesel. Meanwhile the market for CNG-powered buses in North America is set to grow as a result of more stringent emissions standards. Engine manufacturer Cummins is expecting to have around 500 CNG buses in operation by the end of 1994, using conversions of existing diesel engines. Major users include Texas, Florida, New York, California and Ontario (Carter, 1993).

Moderate reductions in CO_2 emissions can be achieved by switching from petroleum to LNG or CNG, and supplies of the fuel are comparable with those of petroleum fuels. However, there is a danger that the modest benefit could be cancelled out by emissions of methane, itself a greenhouse gas, through leakages from methane processing plants and distribution networks. Chapter 2 shows that molecules of methane are some 25 times more powerful as radiative forcing agents than those of CO_2. It has been suggested that emissions of methane could offset the savings in CO_2 resulting from the use of natural gas, and render the fuel worthless as a means of reducing greenhouse emissions. Mills et al (1991) describe the effect thus:

> ... if only CO_2 is counted, compressed natural gas (CNG) automobiles appear 'better' than gasoline-fuelled cars. However, total greenhouse gas emissions are in fact greater for CNG automobiles after including the CO_2 releases from fuel production and related methane and N_2O emissions.

Storage is also a problem for CNG, as the compressed gas is six times less dense than conventional petroleum fuel and requires large, heavy cylinders to be carried aboard the vehicle.

Because of the uncertainty surrounding the scale of methane leakages, the value of natural gas as part of a strategy to reduce CO_2 emissions remains unproven. Furthermore, the longer-term supply of natural gas at competitive prices is subject to some doubt. It has been estimated that known reserves of natural gas will be depleted within 60 years if present consumption rates continue. Natural gas cannot therefore be viewed as a truly long-term option for transport fuel.

The cost of retrofitting a vehicle to run on CNG is considerable. However, purpose-built CNG cars would be unlikely to cost much more than petrol-engined vehicles. Refuelling could take place using the public gas supply, subject to limits on the amount of impurities in the fuel. A critical obstacle to the introduction of natural gas is the limited availability of the fuel, which requires vehicles to be able to run on petrol too. So-called 'dual-fuel' engines fall well below the optimum efficiency of dedicated natural gas engines (Dallemagne, 1990). Overall, natural gas appears to have little to offer as part of an overall greenhouse gas reduction strategy.

Of more value in reducing CO_2 emissions is a fuel closely resembling natural gas, which can be manufactured from biomass such as wood chips, using a thermochemical gasification process. The gasifier uses heat to convert woody biomass into distillates, and can be carried aboard the vehicle. Traditionally, gasifiers have been most commonly used as stationary sources of power in remote areas. As with other biofuels, net emissions of CO_2 from gasified biomass can in principle be reduced to almost zero, if the gasification process is powered by some of the energy that it releases.

Liquid petroleum gas (LPG)

Liquid petroleum gas consists of propane (C_3H_8) and butane (C_4H_{10}), and is a by-product of the drilling and distillation of crude oil. At present around 2 per cent of cars in Europe are powered by LPG, mostly in the Netherlands where the fuel is relatively abundant. Emissions of CO_2 from LPG engines are lower than those from conventional petrol engines, and LPG is also classified as a 'clean' fuel under California's air quality programme. The fuel has been on trial in converted tourist buses in several English cities, principally as a means of reducing the amount of smoky exhaust produced by older vehicles (Millar, 1993).

Conventional petrol engines can be made to run on LPG by an adjustment to the carburettor, though the benefit of LPG in terms of efficiency and emissions tends only to become significant if the fuel is used in purpose-built engines rather than adapted gasoline engines.

Like natural gas, LPG has a lower calorific value than petrol. This means that a larger fuel tank is required in which to store the pressurized fuel, which may cause design problems in smaller vehicles. This is particularly the case if the vehicle is fitted with both petrol and LPG tanks. In addition, the efficient design of LPG engines is at present hampered by the need to run on both LPG and petrol.

Since LPG is a by-product of petroleum production, its supply is limited and depends on the production of conventional motor fuels. Martin and Michaelis (1992) estimate that the supply of LPG could double if the use of 'reformulated gasoline' – from which traces of butane have been separated – were to become widespread. But they hold the view that 'supplies of LPG are not sufficient to justify the development of a larger market in LPG-fuelled vehicles'. In summary, therefore, LPG could be useful as part of a CO_2 reduction strategy, but not on a large scale.

Alternative heat engines

So far, the options for alternative combustion fuels have all been viewed in the context of internal combustion (IC) engines. However, a variety of alternative engine types exist, which offer a potential for reducing CO_2 emissions. The three designs examined here are Stirling engines, gas turbines, and Rankine engines. All three operate using continuous combustion, rather than the intermittent combustion of spark and diesel engines.

Stirling engines

These operate by forcing a gas between hot and cold spaces, and using the expansive work to move a piston. The working fluid is enclosed in a sealed system, with combustion taking place externally. This means that a wide variety of fuels can be used, offering the potential for a broad range of biofuels in solid, liquid or gaseous state. Most of the research relating to automotive Stirling engines has been applied to trucks, buses and large cars. Francis and Woollacott (1981) point to the practical difficulties associated with Stirling engines, but conclude that the Stirling has an advantage 'in terms of noise, vibration, and possibly in emissions'. They also point to 'the long-term importance of fuel flexibility which is greater in the case of the Stirling than for any other heat engine which is a candidate for road vehicles'.

Gas turbines

These, a form of the Brayton engine, make use of a working fluid which is exhausted after each cycle. Although not yet applied to production vehicles, gas turbines have found limited use in racing cars, such as the Rover-BRM which won in its class at Le Mans in 1965. Air is compressed and passed into the continuous combustion chamber where it is heated by the burning fuel. The hot gases then expand to ambient pressure, turning a turbine wheel, and are expelled to the atmosphere.

Gas turbines are associated with a degree of fuel flexibility as well as a long service life. Engine efficiency is optimized at maximum load, which is associated with very high rotational speeds. For this reason,

Dallemagne (1990) suggests that the most suitable application might be in a hybrid vehicle, in which the turbine is operated occasionally for the purposes of recharging the batteries or supplying additional motive power to the vehicle. In general, the application of gas turbines is technically simpler and more energy efficient in larger vehicles such as trucks and buses than it is in typically sized cars.

Rankine engines

These are best known in the form of the steam engine, which was used in commercial railways as recently as the 1960s. Steam cars were in use from around 1900, but were rapidly superseded by internal combustion engines whose efficiency and power-to-weight ratio were higher. The working fluid in a Rankine engine, usually water, alternates between liquid and gaseous states. Efficiency is poor compared with that of other heat engines, but Rankines can be operated on a broad range of fuels. Francis and Woollacott (1981) hold the view that '... it is widely accepted that the Rankine engine is not a serious prospect for road vehicles'.

It appears unlikely that alternative heat engines will offer significant energy savings over, for example, advanced diesel engines. However, their value in reducing CO_2 emissions lies in their fuel flexibility, which allows biofuels to be used, with potentially much lower net CO_2 emissions. On this basis it would appear that the greatest potential for reducing CO_2 emissions from heat engines lies with the Stirling engine.

In addition to the three alternatives considered above, there is also some potential for novel configurations of the conventional petrol engine. Two-stroke engines, smaller and lighter than conventional four-strokes, offer higher levels of fuel economy, by virtue of the direct injection of fuel into the combustion chamber. Advances in fuel injection and engine management have made two-strokes a viable proposition for production cars (Dymock, 1991), though they have been used for many years in the Trabant cars that were once ubiquitous in the former East Germany. A design developed by Ricardo is claimed to offer a power density as high as 90 kW per litre of engine capacity – twice that of a conventional four-stroke engine with four valves per cylinder (Martin and Michaelis, 1992).

Catalytic ignition engines make use of a platinum catalyst to ignite the fuel–air mixture, rather than using spark or compression ignition. NOx emissions tend to be less, by virtue of the lower combustion temperature. This type of engine is still at an early stage of development (Griffiths, 1990). Other innovative designs include the 'quadratic engine' developed by Supertron of London and the Collins 'scotch yoke' engine (McLain, 1990; Fishlock, 1991).

Prospects for alternative rail transport fuels

As a growing proportion of rail operations are powered by electricity, the option of alternative fuels in railways is largely an issue of finding alternative means of generating electricity. The prospects for alternative sources of electricity have already been outlined in connection with electric road vehicles. At present, electric railways draw their power from the national electricity supply, and the electricity is predominantly coal derived. The extent to which renewable energy sources could be developed in the electricity supply industry is beyond the scope of this book, but it should be emphasized that a switch away from coal and towards non-fossil energy sources would have a valuable effect in terms of reducing CO_2 emissions from rail travel.

Alternative fuels may also be developed for non-electric rail operations. A number of fuels are available as a substitute for diesel in locomotives, including natural gas, LPG, hydrogen and alcohols. The technical difficulties associated with developing these fuels for rail use would be fewer than for road transport, since there is less constraint on physical space. In addition, the refuelling regime would present fewer problems, since rail journeys tend to be regular and highly scheduled.

But at present there is no indication that alternative fuels will be developed for rail use, for the following reasons:

- The capital required for an investment in alternative fuels is generally not justifiable alongside more pressing needs such as updating stock, renewing infrastructure and improving safety.
- The main motivation for saving energy in rail operations is economic, and the most cost-effective reductions in specific energy consumption can be achieved through technical refinements in materials, aerodynamics and so on, rather than by switching to new fuels.
- From the perspective of national CO_2 emissions, rail travel represents only around 3 per cent of the total output. A reduction in the carbon-intensity of rail transport would have a minor effect on overall emissions relative to the cost involved, even if the volume of rail travel were to increase substantially.

The most promising prospects for less carbon-intensive rail fuels would therefore appear to lie within the electricity supply industry. The current rail stock consists of a mixture of diesel and electric vehicles, and at present there is little or no reduction in CO_2 emissions associated with a switch from diesel to electric traction. But if the electricity supply industry were to become substantially less carbon intensive, electrification would offer significant reductions in emissions of CO_2.

Prospects for alternative air transport fuels

In principle, aircraft can be operated on a variety of fuels, and a number of them have already been used successfully in prototypes. However, there are technical difficulties associated with alternative aviation fuels, and these impose severe limits on the commercial viability of these fuels (Momenthy, 1991). As with road transport, alternative fuels are only considered useful for the purposes of this book if the overall level of CO_2 and other greenhouse emissions is lower than that from conventional-fuelled vehicles. On this basis, the options available for aircraft are:

- hydrogen, produced from nuclear or renewable sources of energy;
- alcohols, derived from renewable biomass; and
- natural gas.

One of the most serious obstacles to alternative aviation fuels is the problem of fuel storage. Hydrogen, for example, has a high energy content per unit mass, but a low energy content per unit volume, and would require over four times the volume of an equivalent quantity of conventional jet fuel. Methanol has both a large mass and a large volume relative to conventional fuel for an equivalent quantity of energy. Given that commercial aircraft already carry very large quantities of fuel at take-off, it is clear that any additional demands on weight or storage space would be difficult to satisfy. As in the case of road transport, natural gas is of questionable benefit in reducing greenhouse emissions from aircraft, because significant amounts of methane are likely to be released into the atmosphere during processing and refuelling.

The technical difficulties associated with alternative aviation fuels are formidable, particularly for the options that are less carbon intensive than conventional jet fuel. The development of alternative fuels will be more straightforward in land-borne transport systems, which account for the majority of travel-related emissions. Momenthy (1991) has examined in detail the technical challenges of alternative aircraft fuels, and concluded that, in the aviation sector, 'efficiency improvement is the technically and economically best way to control the production of CO_2'.

Alternative fuels: the scope for reducing CO_2 emissions

A study undertaken in the US has compared the greenhouse emissions of alternative-fuelled vehicles with those of conventional vehicles, using various assumptions about the source from which each fuel is derived. The analysis takes into account emissions not only of CO_2 but also of methane and nitrous oxide. The results are reproduced in Table 6.4, and can be used to ascertain which alternative fuel options could be of use as part of a strategy for reducing greenhouse emissions from personal travel.

A vital caveat accompanying results such as these is that they do not always compare like with like. The superior emissions characteristics of a vehicle running on alternative fuels relative to one running on petroleum often has as much to do with the design of the vehicles as it does with the nature of the fuels being used. In particular, the performance of an alternative-fuelled vehicle (in terms of power and

Table 6.4 Relative greenhouse emissions of different energy sources relative to petrol engines in the USA

Fuel feedstock	% change
Electric vehicles, nonfossil electricity	−100
Hydrogen from nonfossil electrolysis	−100
Natural gas/methanol from biomass	−100
CNG from natural gas	−19
Electric vehicles with natural gas plants	−16
LNG from natural gas	−15
Methanol from natural gas	−3
Electric vehicles from current power mix	−1
Petrol and diesel from crude oil	0
Electric vehicles with new coal plant	+26
Metal hydride from coal	+100
Liquid hydrogen from coal	+143

Source Sperling and DeLuchi, 1989

body weight) is often considerably less than that of the conventional IC-engined vehicle against which it is being compared. For example, the General Motors Impact electric car, described earlier, is built from lightweight materials and contains a number of other energy-saving features such as high-pressure tyres and advanced aerodynamics. If these features were to be applied to a small, low-powered, IC-engined vehicle, the resulting fuel economy would be extremely high. Improving the fuel economy of the IC-engined fleet, using technologies such as these, would reduce the benefit offered by a large-scale switch to alternative fuels. The point is also made by Cragg (1992):

> Too often in the debate about contrasting emissions for different types of fuel, the comparisons are not made between similar designs...The acid test, when faced by comparisons between emissions statistics for some alternative fuelled vehicles, is to ask what size of gasoline-driven engine would be needed to power the same vehicle. To return to the Australian example [of a solar-powered car capable of 40 miles per hour], it would be possible to cross the outback at an average speed of 40mph, powered by a lawnmower engine, if the same 'hull' were used as for solar vehicles. The net saving in emissions is thus considerably less than might be expected when comparing solar designs to a Mercedes-Benz. Indeed, if the American public suddenly decided to make their lawnmowers more aerodynamically sound, remove the rotary cutters and use them to go to work, emissions would be slashed.

This tongue-in-cheek illustration reflects a crucially important point. The assessment of alternative fuels made by Sperling and DeLuchi, and summarized in Table 6.4, assumes for a conventional car the current level of fuel economy in the USA, and not the maximum that is technically feasible. In Europe, where cars are on average more economical than in the USA, the benefits of alternative fuels would be less pronounced than Table 6.4 suggests. And if average fuel economy were to be improved still further, through the use of smaller, more efficient cars, the value of switching to new fuels would be even less convincing.

There is a danger that alternative fuels may be regarded by some as a technological panacea for all the environmental ills associated with mass mobility. Although they offer the potential for delivering virtually emission-free travel, this depends critically on the source of primary energy that is used. In particular, renewable energy and biofuels would need to be expanded on a massive scale before hydrogen, electricity and alcohol fuel could contribute significantly to reducing CO_2 emissions. Alternative fuels should therefore be regarded as a long-term option for controlling greenhouse emissions.

With this cautionary note in mind, the following options for alternative surface transport fuels could be considered viable as part of a strategy for reducing CO_2 emissions:

■ Electricity, based on natural gas or renewable energy.
■ Hydrogen (either burned in internal combustion engines or used in fuel cells), with the gas produced electrolytically. As with electric vehicles, the electricity for this process would need to be based on natural gas or renewable energy.
■ Alcohols, oils and gases derived from biomass (fuel crops and organic waste).

Given that the majority of CO_2 from personal travel is produced by private cars, an investment in alternative fuels will yield greatest benefit if applied to this sector. Unfortunately, however, the technical obstacles to alternative fuels are much less severe in buses, coaches and trains than they are in private cars, for reasons associated with vehicle size, predictability of usage, and refuelling (cars require many more refuelling points than other modes of transport). It may therefore be feasible for these modes to adopt alternative fuels too. However, technical difficulties are likely to rule out the use of alternative fuels for air travel.

Dallemagne (1990) has performed a costing exercise for all the alternative fuel options available, based on the year 2010 and assuming all the vehicle's lifetime costs. His estimates are reproduced in Table 6.5. The figures suggest that none of the alternative fuel options considered by this study is likely to cost much more than petrol by the year 2010, given an increase in crude oil prices. Over the timescale of this study, all the options identified here are considered economically viable, though the availability of energy supplies may be limited.

Table 6.5 Cost estimates for alternative fuel options in 2010

Fuel option	Total cost (ECU$_{1990}$ per 100 km)
Petrol	15.9–21.0
Diesel	12.3–15.8
Sodium-sulphur battery	9.9–20.2
Liquefied solar hydrogen	16.5–30.9
Metal hydride (solar)	16.1–27.9
Methanol	14.3–17.6
Compressed natural gas	11.7–16.8

Source Dallemagne, 1990

The role of regulations in promoting alternative fuels

The development of alternative fuels in the USA has been stimulated largely by government regulations, in response to air quality concerns. A number of alternative fuel programmes have been mandated by the federal government's 1990 Clean Air Act Amendments, whilst California's standards require a proportion of vehicles to be 'low-emissions vehicles' by 2000 (see below). The current competition between manufacturers to develop a practical electric vehicle would have been unlikely to take root in a 'business-as-usual' market for cars. The cost of investing in new technology, together with the negligible demand for alternative fuels, would not have allowed car manufacturers to diversify into new forms of propulsion.

The same effect can be found in Europe. Before EC legislation was passed requiring cars to be fitted with catalytic converters, car manufacturers in Britain were reluctant to cater for the new pollution control technology. The prospects for alternative fuels can be viewed from a similar perspective: government mandates aimed at promoting alternatives to petroleum would be likely to stimulate a new market for alternative-fuelled vehicles. Until then, most of the alternative fuel options will continue to be subject to a closed loop: the fuels are not available to the public, so interest is lacking; but the fuels will only become available when demand for the fuels becomes significant. Government regulations can be effective as a means of breaking out of this loop.

The role of market mechanisms in promoting alternative fuels

At present, alternative fuels cannot be justified on conventional economic grounds. Cheap petroleum and an absence of pollution taxes or regulations mean that car manufacturers have little incentive to develop and market vehicles running on 'cleaner' fuels. However, market forces could be exploited to promote alternative fuels if the necessary changes were made to the structure of the market. In particular a carbon tax, as described earlier, would effectively raise the price of fossil-based fuels and make alternative energy sources more attractive to consumers. Crucially, it would also exclude the use of alternative fuels that have no value in reducing CO_2 emissions, such as hydrogen derived from coal.

As an alternative to new taxes, the existing taxation system might be adjusted to recognize the value of 'cleaner' vehicles. Reduced rates of taxation for particular vehicle types or fuels could be effective in promoting sales of less carbon-intensive products.[7] But if fiscal incentives are to be used to promote alternative fuels, they clearly need to be based on whole-cycle emissions of greenhouse gases, rather than just emissions from the tailpipe.

A programme known as 'Proalcohol' was established in Brazil in 1975 as part of a drive to promote ethanol as a motor fuel. The national government intervened in several sectors of the economy, providing incentives for research and investment programmes, giving price guarantees to producers, and increasing taxes on motorists. Cars running on petrol alone were banned during the mid 1980s, and high taxes were levied on imported cars. Only by intervening positively in the vehicle market was it possible for the Brazilian government to develop alcohol fuel on a national scale (Weiss, 1990).

Despite recent improvements in the cost-effectiveness of ethanol production, Brazil is still finding it hard to displace petroleum in a climate of falling oil prices. There are fears that ethanol will settle into a role as a petrol additive, rather than a fuel in its own right. All-ethanol cars, which once accounted for 80 per cent of new cars built in Brazil, could soon become a thing of the past. But whatever the future of Brazil's 'Proalcohol' programme, it is estimated that the initiative has saved the country £20 billion in oil imports, and if the programme can survive the short-term uncertainties inherent in Brazil's energy policy, it is likely to find a secure market for the future (Homewood, 1993).

It is also feasible to use tradeable permits as a means of encouraging car manufacturers to sell alternative-fuelled vehicles. Such a scheme is built into the Californian programme of alternative fuels development, which is summarized by Jackson and Schoon (1991):

> From 1998, any company that wants to sell more than 35,000 cars in the state must make sure 2 per cent of them are 'zero-emission vehicles', emitting no pollutants at all. That figure will rise to 10 per cent by 2003....
>
> Companies are not forced to develop a commercially viable zero-emission car. If they think that is impossible, they have a simple option: to sell their own, or someone else's, electric cars at a subsidized price low enough to attract buyers – and to cover the cost of doing so by raising prices on their petrol cars.
>
> There is one further twist. If a company manages to develop an electric car that accounts for more than 2 per cent of its sales, the firm can claim a credit for every extra one it sells. Those credits can then be sold on the open market to other companies to fulfil their quotas.

The scope of technological solutions to curb CO_2 emissions

The above sections have considered technological changes that could be made to the vehicle stock in order to reduce emissions of CO_2 from individual vehicles. Measures to encourage car buyers to choose more economical vehicles have a considerable potential for reducing CO_2 emissions from personal travel. In addition, alternative fuels can also play a part, though in some cases the technology requires further development. In all cases, the low price of petroleum is the greatest immediate obstacle to implementation.

The technological solutions described in this chapter are unlikely to be adopted on a significant scale in a business-as-usual world. Whilst there is a degree of environmental benevolence in the market for cars, there are no indications that this will be sufficient to bring about major changes in the nature of private vehicles. Instead, regulations and market-based instruments will need to be introduced in order to stimulate interest in these technologies. Market measures tend to be more appropriate than regulations since they do not impinge upon freedom of choice, and tend to offer consumers and manufacturers a variety of possible responses.

The SPACE model has shown that 'business as usual', a continuation of present policies, will result in an 80 per cent increase in Britain's CO_2 emissions from personal travel by the year 2025. Two broad classifications have been proposed for policy measures aimed at tackling this growth: Approach A, consisting of technological changes to the hardware and infrastructure of passenger transport; and Approach B, comprising changes to the structure and volume of personal travel. What contribution can technological solutions – Approach A – make to controlling CO_2 emissions? To answer this question, the SPACE model has been used to predict the outcome of a 'technical fixes' scenario for Britain – Scenario 2 – in which vehicle fuel economy and alternative fuels are vigorously promoted, with no change in the growth of travel demand.

An improvement in the specific energy consumption (SEC) of private cars can be achieved through a combination of better vehicle fuel economy and a shift in the vehicle stock towards smaller engines. In addition, a growing proportion of cars may be run on 'greenhouse neutral' energy sources, of the types identified earlier in the chapter.

The volume of travel demand in Scenario 2 is the same as in Scenario 1, as are the relative shares of mileage by different modes. However, the fuel economy of private vehicles progressively improves via the use of market-based incentives. In addition, alternative fuels are assumed to enter the market too, as a result of similar incentives, and supported by further development of the relevant technologies.

As indicated throughout this study, private cars can be considered a special case in terms of fuel economy. Unlike other modes, the average fuel economy of cars is considerably worse than the cost-effective maximum, as buyers have until now tended to rate comfort, size and performance more highly than 'miles per gallon'. By contrast, bus operators, rail companies and airlines tend to use the most economical vehicles that are available to them (within financial constraints), in the interest of minimizing their operating costs.

Scenario 2 assumes the adoption in Britain of a feebate scheme, of the type introduced in Ontario, Canada, described earlier in the chapter. Fees and rebates are added to car sales tax in order to reflect the fuel economy of different models relative to the average. The scenario examines the combined effect of a system of fuel economy labelling plus financial incentives for purchasers of economical vehicles. The revenue raised by the sales tax surcharges would provide funding for the rebates and for the programme's administrative costs.

An illustrative example of such system is given in Table 6.6. In the absence of empirical evidence, the necessary level at which the fees and rebates should be set is difficult to predict with accuracy. Achieving revenue neutrality is doubly difficult, because not only is the effect of incentives on car purchase decisions poorly understood, but any change in purchasing behaviour will shift the necessary balance between rebates and surcharges. In practice, a flexible, 'iterative' methodology is the most useful, in which taxes and subsidies are initially set at an estimated level, and revised annually on the basis of the past year's performance as part of the national budget statement.

Table 6.6 Illustrative example of a feebate scheme for car sales

Car fuel economy (litres per 100 km)		Surcharge or rebate
Over	Not over	
	6.0	£500 rebate
6.0	7.0	£100 rebate
7.0	9.0	£100 surcharge
9.0	12.0	£1,000 surcharge
12.0		£5,000 surcharge

The fuel consumption of a particular make and model of car is clearly defined by the three official test figures published by the UK government. The Transport Research Laboratory (TRL) has derived a formula that relates these test data to real life fuel economy, using a statistical technique (Watson, 1989). This 'converted' figure, or one derived using a similar formula, could be used to provide a single fuel economy value for each car model, to be used in assessing the fuel economy of different models for the purposes of these two policy measures.

The TRL formula for converting official test results to actual, on-road fuel economy is given as:

$$\text{Consumption} = 0.240\,F_u + 0.308\,F_{90} - 5.47\,\text{ENG} + 20.615$$

where 'consumption' is the average on-road fuel consumption in miles per gallon, F_u and F_{90} are the official list consumptions in the urban cycle and at a steady 90 kmh, and ENG is the engine size in litres.[8] The single figure derived in this way could then used to determine the level of the sales tax surcharge or rebate.

Fuel economy labelling for cars, together with a change in the emphasis of car advertisements away from performance and towards economy, is assumed to take place as a direct result of the fuel economy incentives introduced in this scenario. However, labelling would need to be standardized by means of government regulations, to allow the car buyer to make valid comparisons between models. It is therefore proposed in Scenario 2 that a standard method of calculating a single fuel economy figure be devised, based on the official data that are available for all cars. All new cars would be required to display a notice indicating the official fuel economy value, as well as the fee or rebate payable by the customer at the time of purchase.

A strategy for promoting vehicle fuel economy would have two effects on the car stock. Firstly, purchasing patterns would change: some purchasers would be willing to pay more for their vehicles, whilst others would shift to more economical cars. Overall there would be a net increase in the average fuel economy of new cars, and in the fraction of cars that are diesels. And secondly, the fuel economy of a particular engine size would improve as manufacturers placed more emphasis on fuel economy and less on performance. The net result would be a gradual increase in the average fuel economy of the nation's car stock, becoming more pronounced as old cars were replaced by new.

The approach adopted in Scenario 2 is a 'target-led' strategy. A target for an annual improvement in new-car fuel economy is agreed, and the feebate scheme adjusted to achieve this goal, based on best estimates of the market's response. In this way the level of fees and rebates can be modified each year in order to match the desired target.

For the purposes of this scenario, an estimate has been made of the change in the car stock that could be brought about by the feebate scheme. This projection is not based on observation, but instead represents an estimate of the changes in purchasing patterns that could feasibly be achieved by a feebate scheme, and is illustrated in Figure 6.2. Cars currently in use are allowed to work through the system to the end of their lives, and the process of altering car-purchasing decisions is assumed to take place gradually, so that the distribution of cars shown at the right-hand side of the graph is eventually reached. The growth in the diesel car population, particularly those of engine capacity 1800 cc and under, is assumed to increase markedly as a result of the fuel economy incentives, whilst within the petrol-engined categories there is a gradual shift towards cars with smaller engine capacities.

As explained above, the fuel economy measures adopted in Scenario 2 would not only change the pattern of engine capacities, but also influence

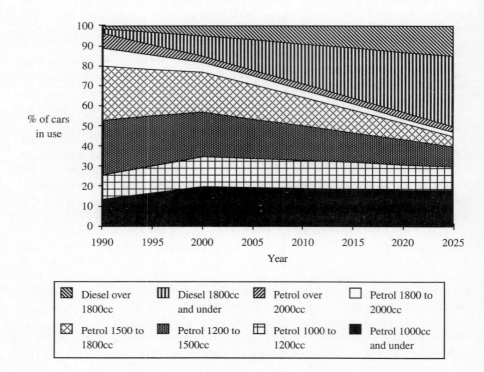

Figure 6.2 Engine capacity distribution resulting from a feebate scheme

the fuel economy of a given engine size. Martin and Shock (1989) have introduced the concept of *benchmark* fuel economy, defined by what is technically possible for a given engine size category. Benchmark fuel consumption represents 'a car design and performance characteristic which is already available in the car market within each category of engine capacity', calculated as 'the 20th percentile of each distribution of overall fuel consumption in each vehicle category' (ibid). Table 6.7 compares average and 'benchmark' fuel economy for the British car stock in 1986. Scenario 2 assumes that in the period to 2005, the market measures outlined above lead to a gradual transition from average fuel economy to benchmark fuel economy, or 'state of the art'.

Based on the level of improvements described above, the target adopted for Scenario 2 is *an annual increase of four per cent in average new-car fuel economy over a period of five years*. This corresponds to an eventual improvement of 20 per cent in the fuel economy of the whole car stock. The car stock is assumed to be replaced completely after 15 years. After 25 years, the improvement rises to 38 per cent, as a result of ongoing improvements in fuel efficiency and further changes in purchasing behaviour.

Table 6.7 Average and 'benchmark' fuel economy in Britain, 1986 (litres per 100 km)

Vehicle category	Average fuel economy	'Benchmark' fuel economy
Petrol-engined		
Up to 1000 cc	7.6	6.2
1001–1200 cc	8.3	6.7
1201–1500 cc	8.9	7.1
1501–1800 cc	9.8	7.4
1801–2000 cc	10.6	7.9
Over 2000 cc	13.7	9.8
Diesel-engined		
Up to 1800 cc	6.4	5.4
Over 1800 cc	7.7	6.6

Source Martin and Shock, 1989

Figure 6.3 illustrates the improvement in fuel economy implied by this target, and compares it with the 'business-as-usual' trend seen in Scenario 1. This is regarded as a realistic estimate of the improvement in car fuel economy that could be brought about using a feebate scheme for car buyers. It is viewed as conservative, reflecting what could be achieved in the real world rather than in optimistic forecasts.[9] As explained above, the increase in fuel economy would take place as a result of a net shift towards smaller engine sizes, plus a movement *within* engine size categories towards 'best available technology'.

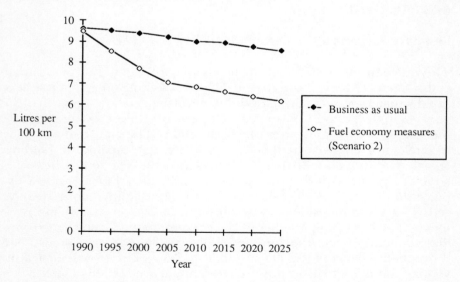

Figure 6.3 Car fuel economy improvements targeted in Scenario 2

As noted earlier in the chapter, improvements in vehicle fuel economy can, in isolation, lead to an increased demand for travel, as the fuel cost per kilometre decreases. Scenario 2 takes account of this effect, using best estimates of the elasticity of traffic level with respect to fuel price (Goodwin, 1988). The result is a marginal increase in car kilometres compared with 'business as usual', which is included in Scenario 2.

An alternative fuels programme

Scenario 2 also includes a programme for the research, development and deployment of alternative-fuelled vehicles as part of a strategy to reduce emissions of CO_2 and other greenhouse gases. The options for such vehicles have been narrowed down to electric and hydrogen vehicles charged using renewable energy, and heat-engined vehicles running on renewable biofuels.

It is unrealistic to suppose that the entire car stock could be replaced with alternative-fuelled vehicles by the year 2025. There are severe technological difficulties associated with such a transformation, not least of which is the availability of sufficient quantities of renewable energy. Instead, it is expected that alternative fuels would be introduced as part of a target-led programme of incentives, coupled with further research and development. For the purposes of Scenario 2, it is assumed that alternative-fuelled vehicles will begin to enter the market in significant numbers in 2005, with 10 per cent of the national car fleet running on 'greenhouse-neutral' sources of energy by the year 2025. The growth in alternative fuels is likely to continue thereafter, beyond the timescale of the SPACE model.

Results of Scenario 2

The outcome of Scenario 2, in terms of passenger transport's total emissions of CO_2, is illustrated in Figure 6.4. Emissions increase less rapidly than they did in Scenario 1, but nevertheless show a growth of 35 per cent in the period 1990 to 2025. Of particular interest is the shape of the graph before and after the year 2005. The figure shows that technological changes to the car stock are not capable of holding emissions of CO_2 down to their 1990 level up to the year 2000 – the target adopted by the British government as part of a Europe-wide CO_2 programme. After 2005, they begin to rise slightly more steeply, reflecting the diminishing returns offered by fuel economy improvements. The increasing number of 'greenhouse-neutral' vehicles to some degree holds down the long-term growth in CO_2 emissions; but throughout the period 2005 to 2025, the ongoing growth in traffic volume continues to force upwards the annual production of CO_2.

These results have important implications for the setting of national 'greenhouse' targets. Scenario 2 shows that 'technical fixes' are not capable of reducing Britain's emissions to the 1990 level by the year 2000, the target adopted by the European Community. In the period 2000 to

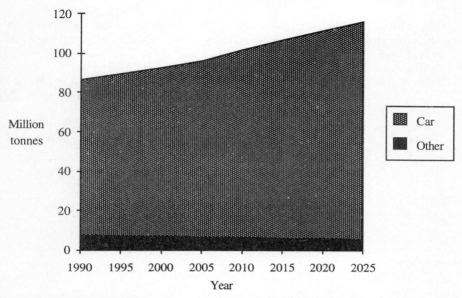

Figure 6.4 Total emissions of CO_2 from personal travel in Scenario 2

2025, the growth in emissions continues as fuel economy improvements reach their technological limit, although the alternative fuels programme is beginning to have an effect. In the timescale of the SPACE model, there is clearly a need for complementary measures to be adopted if CO_2 emissions are to be stabilized or reduced.

The IPCC has estimated that industrialized countries will need to cut their emissions of CO_2 by at least 60 per cent in order to stabilize the atmospheric concentration of this gas. It was suggested earlier that personal travel might be required to achieve a cut of 20 to 25 per cent as part of such an overall strategy. The following chapter considers how additional policy measures, affecting not only the design of vehicles and fuel but the demand for travel itself, might be deployed alongside 'technical fixes', in order to achieve such an objective.

Notes

1. One way of avoiding this effect is raise fuel prices at least by a percentage that corresponds to the rate of improvement in fuel economy.
2. Further savings could be made by enforcing the law on roads with a 60 mph speed limit.
3. The drag coefficient of the Impact is 0.19.
4. It should be noted, however, that nuclear stations use a large amount of fossil fuels in their construction, a consideration that needs to be taken into account in any CO_2 reduction strategy.
5. CO_2 emissions from natural gas-fired power stations will be reduced with the introduction of more efficient combined-cycle generation.
6. When hydrogen is burned in internal combustion engines, some NOx is produced. However, emissions of this gas can be minimized through the use of lean mixtures, as well as various methods of combustion cooling.

7. At present electric vehicles are exempt from Vehicle Excise duty (VED) in Britain. However, this tax constitututes a small fraction of motoring costs (£125 per year for private cars), and the incentive is not, therefore, sufficiently large to stimulate a market for electric vehicles.
8. All consumption figures are in miles per gallon (mpg). The formula applies only to petrol-engined cars, but there is no reason why a similar formula might not also be derived for diesels.
9. Many experts have agreed that the average fuel economy of new cars could be improved by 40 to 50 per cent over the next 10 to 15 years (European Federation for Transport and Environment, 1992).

7

Questioning the Need to Travel

The private motor car has assumed a status beyond its utility value as a means of transport. There is a lack of general acknowledgement or appreciation of the environmental disbenefits associated with increasing levels of car usage and fuel consumption.

Department of the Environment and Department of Transport

The two scenarios examined so far indicate strongly that technological measures alone will be insufficient to curb CO_2 emissions in the light of the enormous expansion in road transport that is forecast to take place in Britain. There is good reason to believe that most other developed countries will find themselves in a similar position – with the possible exception of the USA, where car ownership and vehicle mileage are showing signs of levelling off at a saturation limit.

The clear message, therefore, is that more stringent measures will be required in order to reduce, or even to stabilize, emissions of CO_2 from personal travel. These will need to address the demand for travel itself, in order to dampen the projected increase in road traffic. The aim is twofold: to achieve transfers from car travel to less CO_2-intensive modes such as rail, bus, walking and cycling; and to reduce the absolute growth in travel by all modes. The two objectives are to some extent connected, since the widespread acquisition of cars has tended to be associated with longer, and more frequent, journeys (see Chapter 3).

As with the 'technical fixes' contained in Approach A, measures for altering the nature and volume of travel demand can take the form of regulations, market-based instruments or investment. All three types of policy lever can be used to influence personal travel decisions, with a target of reducing CO_2 emissions.

Historically, three factors have contributed to the present volume of car usage:

■ Rising car ownership leading to a transfer to car from other modes of transport.

- Rising car ownership leading to the generation of additional trips.
- Rising car ownership leading to a lengthening of the average trip (an increasingly important effect).

The contribution of the last two of these factors has increased in importance. In terms of reducing emissions, simple transfers to more energy-efficient modes offer only limited scope, because the historical growth in car use has led to a pattern of travel demand and land use that could not be catered for by other modes. Nevertheless, if public and non-motorized transport were to be expanded to their maximum possible modal share, a significant volume of car travel could be eliminated. In urban areas especially, car trips for shopping, work and leisure purposes could be replaced with walking and cycling, by placing greater emphasis on the use of local facilities.

It is therefore necessary to look beyond simple modal transfers as a means of influencing travel behaviour, and consider ways of reducing travel demand itself. Policies of this type are often termed *transport demand management*. A reduction in travel volume can be achieved by addressing the second and third of the areas identified above.

This chapter will examine policies that could be deployed in order to modify current patterns of travel demand towards a less CO_2-intensive framework, involving both modal transfer and reductions in travel volume. The options below are ordered with reference to the three forms of policy measure introduced above – regulations, market-based incentives and public investment.

The role of regulations in altering travel demand

As indicated in Chapter 1, personal travel should be viewed as a means of gaining access to people and facilities, rather than an activity in itself. Policies aimed at reducing mobility should not, therefore, lead to any loss of amenity, provided that access is maintained or improved. It is no exaggeration to suggest that restraining car traffic, especially in towns and cities, can in many cases lead to an overall improvement in access, by enhancing the efficiency and reliability of buses, and improving the environment for cyclists and pedestrians.

There is considerable potential for reducing travel demand through land-use planning policies. As Chapter 3 has shown, the historical growth in car ownership and personal mobility has led to an increasingly dispersed pattern of land use. Manifestations of this effect include the retailing centres and housing developments found on the outskirts of most major towns and cities. Meanwhile multiple car ownership has allowed a decentralization of population to take place, with an increasing number of households living in remote areas and commuting relatively long distances to work.

Chapter 3 showed that the number of journeys per person in Britain has changed little since 1975, or between different settlement densities.

Owens (1986) suggests that any reduction in overall travel would be brought about principally by the introduction of shorter journeys, rather than by reducing the number of trips. This fits well with the need for equitable solutions to transport problems. A scheme whereby journeys were 'rationed' would raise justifiable objections on the grounds of freedom of access.

Land-use planning

To reduce CO_2 emissions through planning, there are three aspects of land use that need to be addressed: urban density, settlement size, and urban and regional structure. Chapter 3 showed that the demand for travel, particularly by car, could be reduced by moving towards fairly large, high-density settlements, in which there is a high degree of centralization. There may also be benefits associated with a 'polycentric' structure, though the evidence remains patchy.

As Chapter 3 indicated, a fairly strong inverse correlation exists between urban density and personal travel demand. Dense settlements are also better able to support modes of transport other than the car.[1] There is thus considerable scope for reducing emissions of CO_2 from personal travel through planning regulations. Policies might include the introduction of minimum density standards on new development, and the encouragement of development in existing urban areas – perhaps using vacant or derelict land (Departments of the Environment and Transport, 1993).

Settlement size has a less straightforward influence on travel demand, though it is clear that larger settlements, by virtue of being more self-contained, tend to be associated with a lower travel intensity. Planning regulations could usefully be employed to encourage new housing development to take place within existing larger settlements, whilst discouraging the expansion of small, isolated settlements that are poorly served by public transport. Self-containment can also be encouraged by ensuring that any new settlements are located well away from existing large towns and cities (ibid). Such a policy appears to have worked well in the case of the new town of Milton Keynes. Situated approximately midway between London and Birmingham in the Buckinghamshire countryside, the town has a relatively high level of self-containment. A recent travel survey has shown that only 3 per cent of work trips by the city's residents are to London (TEST, 1991).

Finally, urban and regional structure has been shown to have a significant influence on travel behaviour. A detailed study has been undertaken to compare the travel behaviour of residents in Milton Keynes with that of people living in Almere, a comparable new settlement in the Netherlands. Figure 7.1 compares the modal distribution of travel in the two towns, and shows that car travel accounts for the majority of trips in Milton Keynes – 69 per cent, compared with 43 per cent for Almere. Travel by bicycle, on the other hand, is far more popular in Almere than in Milton Keynes: over a quarter of trips in Almere were by bicycle, compared with just 6 per cent in Milton Keynes. The study concludes that '... the difference in modal split can be attributed to the dissimilar land use structures, travel facilities and culture of the two cities' (ibid).

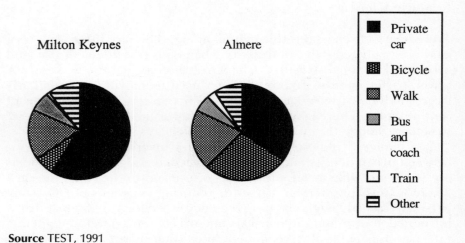

Source TEST, 1991

Figure 7.1 Modal distribution of trips in Milton Keynes and Almere

Meanwhile Steadman and Barrett (1991) have investigated the effect of different land-use structures on transport-related CO_2 emissions. Using a computer model, they compared five different configurations against a 'typical' town of population 72,000. They showed that reductions in energy demand and CO_2 emissions could be achieved by switching to two of the five hypothetical arrangements – a centralized structure, or else one consisting of a number of dispersed 'village' centres surrounding a traditional centre. The authors concluded that savings of 10 to 15 per cent in energy use and CO_2 emissions could be achieved through land-use changes at the city-regional scale over a 25 year period.

By encouraging urban development to take the form of energy-saving designs such as these, national government could make a significant long-term contribution towards reducing energy consumption and CO_2 emissions arising from personal travel. Government regulations could be introduced requiring businesses, retailers, public services and so on to choose locations that can be easily reached without the use of a car. For example, planning permission for retail outlets might be made easier to obtain for city-centre locations than for out-of-town sites that can only be reached by car. Peripheral locations well served by roads could be reserved for goods-intensive activities such as distribution (Departments of the Environment and Transport, 1993).

Planning policies to reduce CO_2 emissions are not restricted to the regional scale. At a local neighbourhood level, there is considerable scope for encouraging the use of local facilities which can be reached on foot or by bicycle. The potential for switching car trips to non-motorized modes of travel cannot be overstated, given that over a quarter of car trips to local centres are less than 1.6 km in length (ibid). Policies might include the promotion of development in the vicinity of neighbourhood centres, as well as an investment in the provision and maintenance of high-quality centres. Access to these facilities by non-motorized modes could be encouraged by improving facilities for pedestrians and cyclists.

As part of a package of measures, therefore, land-use planning to reduce the need for travel, together with an improved availability of public and non-motorized travel facilities, could bring about a substantial change in modal distribution, as well as a reduction in overall travel. At the regional level, centralization encourages public transport as an alternative to the car, while at neighbourhood level, well-designed local facilities can provide a large number of services within walking distance of many people's homes. Meanwhile a number of policies are available for making cycling and walking more attractive. Planning policies such as these may be deployed as a means of reversing the historical trend towards car-dependent, travel-intensive lifestyles.

The Netherlands example

Many of the land-use policies recommended by academics as a means of reducing travel demand are currently being put into practice in the Netherlands, as part of the third National Transport Plan (*Tweed Structuurschema Verkeer en Vervoer*, or SVV2+), which was adopted by the Dutch government in 1990 with the objective of limiting the forecast growth in road traffic. All potential development sites are classified into one of three categories – A, B or C – according to their location and accessibility by public transport. Different types of development are then allocated to the three classes, according to their size, number of employees, and relative dependence upon road transport (Sturt, 1992). The classifications are as follows:

- *A locations*. Public transport locations situated in city centres close by the central station, and with limited carparking possibilities. The businesses and developments eligible for these sites are those that have many employees/visitors and are largely independent of goods vehicles.
- *B locations*. Locations with good public transport and good road links, often adjacent to the suburban railway station or some other mode of high-quality public transport. There is relatively little space for long-term parking. Appropriate developments for these locations are those that have moderate job/visitor intensities which at the same time have considerable dependence on road (freight) vehicle access.
- *C locations*. Locations on the urban periphery with direct links to the motorway system but less well served by public transport. Activities

which are largely road dependent and which involve relatively little journey-to-work traffic are eligible for these sites.

Thus, for example, C locations are reserved for industrial and distribution purposes, and may not be used for office developments. It is expected that Rotterdam will have three A locations near the central railway station, and four B locations at regional nodes on the public transport network, with further sites planned around new metro stations (ibid).

The scope for reducing CO_2 emissions through planning appears to be particularly significant in the cities of North America and Australia, where there is less constraint imposed by an established urban structure. Historic cities in Europe, on the other hand, tend to have developed in a fairly dense, centralized structure, so many of the high-density characteristics associated with low transport demand are already in place.

There are, however, caveats involved when recommending that businesses move to more centralized patterns of development. Firstly, there is a danger that a particular region or country could be placed at a commercial disadvantage relative to its neighbours by adopting less transport-intensive planning principles. For example, at the beginning of 1992 the multinational electronics company Toshiba abandoned plans to locate its European head office in the Netherlands because of what it saw as excessive restrictions on development (ibid). Therefore unless development control measures are established on a universal basis, it is possible that individual governments will be placed at an economic disadvantage by adopting them. Secondly, the classical inverse relationship between land-use density and energy consumption does not always hold. London, for example, comprises a massive employment centre surrounded by an extensive region from which workers commute, often over long distances.[2]

Thirdly, attracting more commuters into the centre of a city could, in some instances, lead to an increase in CO_2 emissions, particularly if there is an increase in the amount of traffic congestion. Figure 3.12 has shown that car fuel economy is generally poor in low-speed, congested conditions.

Moreover, land-use policy alone is not necessarily sufficient to influence travel volume. Steadman and Barrett (1991) point out that:

> Time budget studies show that people with higher incomes and better access to cars spend roughly the same total amount of time on travel per week as poorer people without cars. Thus people do not spend money to save time on travel overall; rather they devote the same time to travelling further by faster modes. All this indicates that a reduction in car travel requires behavioural and attitudinal changes, as well as land use policies.

The 'fixed time budget' concept is supported by Whitelegg (1993), who points to 'a very rough correspondence in the amount of time each of us devotes to travel regardless of how far or how fast we travel'. If the hypothesis is correct, the implications are profound. It suggests that even if a person is relieved of the *need* to travel – perhaps by locating their home adjacent to the place of work – then they will find other ways of filling the daily 'time budget' allocated for travel, with other types of trip. The key, therefore, is to find less polluting ways with which to fill the time budget. This generally implies shorter, slower journeys, undertaken using public or non-motorized travel rather than car, air or high-speed train. Land-use policies should therefore be used in conjunction with measures aimed at influencing individual travel choices, if they are to be successful in reducing CO_2 emissions.

Parking

A further area in which regulations can be used to discourage car travel relates to the provision of parking. The availability and cost of carparking facilities in urban centres is a key determinant of the amount of traffic entering the centre, and it has been shown that the decision to use a car for the journey to work is strongly influenced by the availability or otherwise of a carparking space at the workplace. Regulations can be used to determine the amount of parking provision, the level of enforcement, and pricing policies. Used in conjunction with improvements in public transport, parking regulations can be a valuable tool in shifting journeys from the car to less polluting modes of travel.

Parking policy does, however, have its drawbacks as a means of reducing the demand for car travel. If the cost of parking is increased, some drivers may actually travel further in order to find facilities where parking is cheaper. In cases where parking capacity is constrained, they may drive around looking for a space. Neither outcome is desirable in terms of reducing fuel consumption or CO_2 emissions. Furthermore, although local authorities in Britain and Germany are beginning to recognize the importance of parking policy in influencing travel demand, there is in many cases a conflict of interest between the need to deter car traffic in town centres and the desire to maximize retail turnover in the face of competition from towns in the same area. Whilst there appears to be a distinct movement towards the elimination of all-day parking in towns, local authorities are reluctant to introduce policies which would deter short-stay visitors such as shoppers and tourists.

Alongside reductions in the availability of carparking space in cities, park-and-ride schemes offer considerable scope for reducing car traffic in city centres. These involve the provision of carparking space on the outskirts of town, and a frequent, cheap bus service to carry visitors into the centre. Alternatively, such facilities may be located adjacent to a railway station, as in the case of the 'parkway' stations that are popular for commuters in Britain. In terms of CO_2 emissions, however, park-and-ride should be treated with caution. In many cases it may only affect a small part of a car journey, and it is not yet clear to what extent park-and-ride actually encourages the use of cars for visits to urban centres. By making travel into town centres quicker and more convenient, park-and-ride facilities may encourage commuters to live further from their workplace, and to consume more energy in their daily travel to and from work. It may also encourage the use of cars among people who would otherwise have made the journey wholly by public transport. Nevertheless, by reducing the amount of car traffic in urban centres, park-and-ride tends to make town centres more pleasant places to spend time in, thereby improving the safety and convenience of walking and cycling.

Work-related travel

Travel to and from the workplace can be influenced by means of employer-based schemes, mandated by national or local government. Work-related travel accounts for 27 per cent of all car trips in Britain (Department of Transport, 1988), and policies aimed at curbing the use of cars for journeys to and from work therefore have considerable potential for reducing fuel consumption and greenhouse emissions. Given that work trips are generally regular and scheduled, the opportunities for transferring to public transport are particularly good. While the private car is often defended on the grounds of the unrivalled freedom and flexibility that it offers to the user, it is difficult to apply this argument to the average trip to and from work, which generally takes place along the same route and at the same time every day.

In the USA, air pollution and, to a lesser extent, traffic congestion have reached crisis levels in several cities. Recognizing the inherent limitations of the catalytic converter as a remedy, politicians have begun to look beyond technology as a solution to air quality problems, and to consider the potential of transportation demand management (TDM). Several pieces of legislation have been passed with the aim of reducing car mileage by promoting increased load factors, modal transfer, park-and-ride schemes and teleworking. These include:

- *The federal Clean Air Act Amendments (CAAA) of 1990*, whereby employers with 100 or more staff are required to increase vehicle occupancy by at least 25 per cent for journeys to work.
- *The Congestion Management Program*, which requires local governments to assess the transport implications of new developments, and if necessary take action to reduce traffic levels (via fees, park-and-ride, public transport provision and so on).
- *The California Clean Air Act of 1988*, which requires average car occupancy in the state to be raised to 1.5 by 1999, and aims to establish a cost-effective list of priorities for air pollution control measures.

- *The South Coast Air Quality Management District (SCAQMD) Regulation XV*, aimed at reducing the number of vehicle miles in the morning peak period: employers of more than 100 people in southern California must submit a plan for reducing single-occupant car travel in commuting trips.

A survey known as *State of the Commute* has monitored commuting patterns in several areas of California since 1989, and found a general increase in car pooling, van pooling and travel by non-motorized modes (see Table 7.1). In response to the legislation, a large number of companies have established pilot schemes with positive results (Collier, 1991). Although the historical data do not extend earlier than 1989, they demonstrate a definite trend away from single-occupant car trips and towards shared cars and public transport. A separate survey of work-places affected by Regulation XV. has demonstrated a significant increase in car pooling over the last two years, with a resulting improvement in average car occupancy. The researchers are optimistic that the trend can be sustained, and that vehicle mileage and air pollution will decline as a result (Giuliano and Wachs, 1993).

Table 7.1 Travel modes of commuters in southern California

| Mode | Percentage | | |
	1989	1990	1991
Drive alone	83	79	78
Car pool	11	14	14
Van pool	0	1	1
Bicycle	2	1	1
Motorcycle	1	0	0
Public bus	2	4	5
Private bus	0	1	0
Walk or jog	1	1	1
Commuter rail	n/a	n/a	0

Note n/a = not asked
Source Collier, 1991

Regulation XV has prompted renewed interest in the practice of *telecommuting* or *teleworking* – working from home using an electronic interface. A number of employers in California have introduced telecommuting schemes as part of their strategy for reducing peak-hour commuting. According to British Telecom, a 10 km car trip uses the same amount of energy as a 21-hour telephone call.[3] It estimates that as many as 15 per cent of employees could soon be spending part of their week working from home. Added benefits would include reduced fuel costs for industry, increased productivity, and a fall in road accidents, BT believes (British Telecom, 1991).

More than 50 'telecottages' – local centres from which people can telecommute – are now operating in remote parts of Britain, each typically used by 40 people. The newly formed Telecottage Association estimates that around 10,000 people in Britain could be telecommuting in a few years' time (Watts, 1993).

A survey of telecommuters in the USA by Pendyala et al (1991) suggests that it is not only journeys to and from work that are affected by teleworking. The average journey for non-work purposes was found to be significantly shorter for part-time telecommuters than that of non-telecommuters – even on 'away' days when the normal journey to work was undertaken.

Regulations aimed at stimulating employer-based travel schemes would be likely to have positive results in terms of reducing car usage for work purposes. However, at the moment there are a number of incentives for commuters to continue travelling by car, particularly in Britain, such as the company car subsidy and the provision of free fuel and parking. It would not therefore be sensible to embark upon trip reduction strategies, such as those described above, until the current 'incentives to drive' had been eliminated.

The role of market-based instruments in altering travel demand

In addition to the regulatory measures described above, a number of market-based policies could be put in place as a means of dampening the projected growth in road traffic. In particular, the way in which transport is taxed has a significant influence on travel volume and modal choice. The subsidization of company motoring by government tends to encourage the use of cars, partly by discouraging the use of other modes and partly by increasing the overall level of travel. In Britain, many company motorists receive some subsidy for private and commuting journeys as well as for business travel. By rationalizing the taxation of company motoring, the incentive to drive could be eliminated.

Taxes

A number of the fiscal measures designed to promote fuel economy, discussed in Chapter 6, could also be used to influence modal choice. For example, relating taxation to car use rather than car ownership – perhaps through road pricing or additional fuel taxes – would increase the marginal cost of using a car, leading to a degree of modal transfer. However, it would be a mistake to assume that car ownership has no effect on the overall amount of travel undertaken. Measures designed to deter the use of cars can have only a limited effect in terms of transferring trips to public transport, because the act of acquiring a car leads to the generation of new trips that were not previously made. By their very nature these trips, which have been made possible only through the acquisition of a car, cannot be transferred back to public transport (Wootton, 1993).

There is a strong case, therefore, for maintaining car purchase taxes at a high level in relation to incomes, rather than replacing them altogether with taxes on travel. A statistical analysis of energy use in transport across the whole OECD has confirmed that car ownership is a major determinant of fuel consumption per capita in OECD countries. Meanwhile the level of public transport use is the least significant influence on fuel consumption (Schipper et al, 1991). Findings such as these serve to reinforce the argument that growth in car traffic can only be effectively contained through policies aimed at discouraging car ownership, as well as use.

Fuel pricing

In Chapter 6, fuel pricing was proposed as a possible means of promoting fuel economy in new cars. But taxes on motor fuel can also be used as an instrument for reducing car usage, by increasing the marginal cost of each journey undertaken. Evidence to date suggests that traffic levels tend to show a short-term dip as a result of increases in fuel price, but that they recover again in the longer term as motorists change to more economical cars. Travel demand is not very responsive to increases in fuel price: in the face of rising fuel prices, the long-term response of car users is to switch to a more economical car in order to restore the level of mobility to which they are accustomed. Fuel pricing is not, therefore, an ideal policy for influencing the demand for car travel (Dix and Goodwin, 1982), but is nevertheless useful as a means of promoting energy conservation and curbing CO_2 emissions.

Business location

Another, more indirect form of taxation on mobility takes the form of business taxes, paid by companies on their premises. It has already been shown that changes in the location of businesses and other commercial activities can have a substantial effect in terms of reducing travel demand. The use of variable rates of taxation to encourage businesses to locate their activities in places accessible by fuel-efficient and non-motorized modes would have a substantial, though gradual, effect on land-use patterns. Businesses could be encouraged to select locations near town centres or those well served by public transport, through the introduction of an access-weighted system of local business taxes.

The Metropolitan Transport Research Unit (1990) has proposed a method of measuring accessibility, based on 'accessibility maps' for particular urban areas. These incorporate, among other measures, walking times to the nearest bus or rail stop. The maps indicate by means of 'accessibility contours' the availability of public transport services to residents of a particular area. Such a definition of accessibility could be used as the basis of a local tax on businesses, levied in proportion to their relative proximity to customers and employees. Reduced taxes could also be made available to businesses that adopted employee travel programmes designed to reduce car use.

The value of such measures as these remains speculative, and further research would have to be undertaken in order to ensure the effectiveness of measures to promote access by reducing the need to travel. It is, however, a crucial area for long-term policy development.

Parking

As discussed earlier, the level of car use in towns and cities is closely related to the provision of parking in the urban centre. The use of parking fees is an effective market-based tool for influencing the amount of travel into towns. Whitelegg (1990a) suggests that car parking spaces be allocated a 'rental' charge in order to promote the use of other travel modes for journeys into cities. He takes the view that charges should be set in a 'target-led' manner, rather than as the result of an abstract consideration of costs and benefits. 'Car parking in cities should be charged at whatever rate reduces its severity in a direct relationship with public transport charges and measures to improve the attractiveness of public transport' (ibid).

Road pricing

Another option for reducing the demand for car travel through the use of market forces is road pricing. In general, road pricing requires the motorist to pay at the point of use for access to the road network. In urban areas, drivers might typically be charged for entering a certain area (usually a congested one), often depending on the time of day. *Congestion charging*, as it is often known, has been widely advocated as a means of reducing urban traffic congestion, although there are other benefits including improved air quality and an enhanced urban environment.

Urban road pricing has in many cases been introduced primarily as a means to generate income for major infrastructure projects such as new motorways. For example, road pricing was introduced in the Norwegian capital, Oslo, in 1990, with the objective of financing a series of new roads in the area. A 'cordon' was set up around the city, and motorists were charged for entering the zone. Payment is made in one of three ways: cash may be handed to a cashier stationed at the toll booth or alternatively deposited in a coin-slot machine; or else the driver may become a subscriber, and pay by means of an electronic 'tag', placed inside the windscreen, which identifies the vehicle. Roadside detectors are used to read the tag, which then debits the appropriate amount from the driver's monthly account. Other cities that have introduced urban pricing as a means of improving the road network or public transport, or both, include Bergen and Trondheim in Norway, Stockholm and Singapore (Lewis, 1993).

The first city in Britain to experiment with urban pricing is Cambridge, where an elaborate system of in-car electronic detectors and 'smartcards' has been designed to charge motorists whenever the car encounters congested conditions. It has been estimated that traffic levels in the city would be reduced by 30 per cent in the first year of operation as a direct result of the pricing regime. However, a change of political control at Cambridgeshire County Council in 1993 meant that the scheme will almost certainly be abandoned after the demonstration period is over, in favour of a more public-transport-oriented approach to traffic restraint.

The British government has meanwhile commissioned a three-year study into urban road pricing, based in London, part of which involved a review of road pricing technology by a team of researchers from the University of Newcastle upon Tyne. This concluded that the technology would probably be sufficiently well developed within ten years for a London-wide scheme to be implemented. The report suggested that the obstacles to road pricing are more likely to arise from administrative and political difficulties than from technical ones (Department of Transport, 1993c).

A common criticism of urban road pricing is that it may simply rearrange the existing distribution of car trips (for example by promoting off-peak commuting) without affecting the overall modal distribution. There is a danger that road pricing will simply restrict access to town centres to drivers who can afford to pay for it (Whitelegg, 1991). However, the prospects for road pricing as a tool for discouraging car use could be vastly improved if it were to be introduced in conjunction with complementary measures aimed at promoting the use of alternative modes, such as investment in and subsidization of bus and rail services. Area licensing could also be linked to measures to promote car sharing, such as the dedicated car-pool lanes that are currently operating in a number of American cities.

Area licensing

A number of area licensing programmes introduced world wide have demonstrated significant benefits, including those in Singapore, Milan, Stockholm, and the Randstad in Holland (MTRU, 1991). Accompanied by improvements in the quality and capacity of alternative modes, area licensing can substantially reduce the number of cars driving into and around urban areas. A large proportion of the journeys affected are daily commuting trips. In most cases, vehicles entering the designated area are required to purchase a permit. The Singapore scheme, established in 1975, requires all vehicles entering the restricted zone to display a windscreen ticket. These are checked visually by police officers stationed at 29 entry points, who also check for vehicles carrying fewer than four passengers – an offence which carries a heavy fine.

In addition, the original programme consisted of improvements to public transport, new traffic regulations and controls on car ownership. Car traffic into the restricted zone showed an immediate reduction of 71 per cent in the period 7.30 to 9.30 am, when licensing was applied. The longer-term effect has been to shift journeys to off-peak hours, and to transfer daily commuting trips to other modes. The modal share of traffic held by cars diminished from 43 to 22 per cent between 1974 and 1988, whilst that of public transport increased from 46 to 63 per cent (ibid).

It should be emphasized that the Singapore area licensing scheme was operated as part of a package of measures, including an investment in new bus and rail facilities and controls on car ownership. An integrated approach such as this goes a long way towards overcoming the potential difficulties of road pricing raised earlier. The positive results achieved in Singapore add weight to the argument that policy measures to reduce travel demand will be most effective if they are introduced as part of a strategic package.

In Milan, access to the city centre by car is not priced, but restricted to residents, businesses and emergency services only. This is, in effect, a regulatory rather than a market-based application of area licensing. Introduced in 1985, the scheme is estimated to have reduced car traffic in the morning peak hours by between 40 and 55 per cent. An estimate of the modal breakdown suggests that 41 per cent of the diverted car traffic has been shifted to public transport, whilst 36 per cent transferred to 'park and walk'. The remainder is accounted for by altered travel times (ibid). Figure 7.2 summarizes the findings.

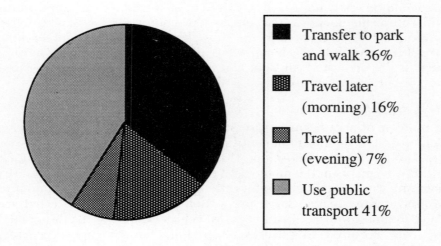

Transfer to park and walk 36%

Travel later (morning) 16%

Travel later (evening) 7%

Use public transport 41%

Source MTRU, 1991

Figure 7.2 Modal transfers resulting from the Milan scheme

Car-free cities

A number of cities in Europe, such as Aachen, Lübeck and Frankfurt, are currently looking beyond area licensing and considering the exclusion of cars altogether. The elimination of cars from city centres is becoming increasingly widespread in Europe, and the 'car-free cities' concept is largely a question of extending the exclusion zone to the entire city area. In 1992 the European Commission published a document stating that car-free cities were a viable proposition, and a number of cities have since expressed an interest in the idea. These include Amsterdam, Aosta, Louvain and Naples, as well as York, Bath and Chester (Whitelegg, 1993; *Local Transport Today*, 1992).

Inter-urban roads
On inter-urban roads, road pricing would be more likely to take the form of a fee, paid by motorists for using a particular stretch of road. By paying each time the road is used, the motorist is charged for roadspace in proportion to their use of it. Advances in 'smartcard' technology are making it increasingly possible to charge motorists automatically, without stopping the traffic. For example, electronic charging has recently been introduced on motorways near Milan, using a system developed by GEC, and is also available to users of London's Queen Elizabeth II Bridge as an alternative to the conventional toll booths.

In recent years the idea of toll roads has received renewed attention as a means of funding large roadbuilding projects in Britain. For example, the proposed Birmingham Northern Relief Road, the first privately funded toll road in Britain, has been financed by private interests who anticipate recovering their costs through charges levied on users of the road. However, critics of inter-urban road tolls have warned that large numbers of drivers might be tempted to use local roads instead of motorways and major roads in order to avoid paying the charge.

The use of toll roads offers the potential for discouraging the use of cars as part of a wider transport or environmental strategy. However, it should be noted that most toll road schemes that have been proposed to date, particularly in Britain, have been instigated not with environmental concerns in mind, but rather as a means of raising private capital for roadbuilding projects. Surveys suggest that public support for road tolls could be improved if some of the revenues raised were allocated to the improvement of public transport, rather than simply paying the developers who built the road. In this respect, Goodwin (1991) has proposed a 'Rule of Three' for allocating the proceeds of road pricing, whereby fees would be divided equally between general tax revenue, road construction and maintenance, and public transport.

In summary, market forces, manifested in their various guises, have a considerable potential for curbing traffic demand and promoting transfers to less polluting modes. In many cases, an adjustment of present taxation structures would achieve worthwhile results; beyond this, additional fiscal measures may be introduced. A system of new taxes could be made revenue neutral, either within the transport sector or across the economy as a whole.

Market instruments are more likely to be successful if introduced as part of a *package* of measures, since such a strategy can exploit the synergies that exist between different policy instruments. For example, road pricing is more likely to achieve modal transfers if public transport is being subsidized at the same time; similarly policies to encourage development of town centre sites will be helped greatly if those sites are served by a high-quality public transport network. Finally, evidence from road pricing surveys indicates that public acceptance tends to be greater if policies are perceived as part of a broad strategy with clear goals.

The role of investment in altering travel demand

As well as regulations and market-based incentives, various forms of government investment can be used to modify the demand for travel and to promote transfers from car and air travel to less CO_2-intensive modes. In fact the role of investment is more useful in altering travel demand than it is in promoting 'technical fixes', which have tended to be brought about via regulations and market incentives. In simple terms, investment has a potential for increasing the capacity or attractiveness of various transport modes through capital spending.

In May 1989 the British government announced details of 'a greatly expanded motorway and trunk road programme', based on estimates of considerably increased road traffic up to the year 2025 (Department of Transport, 1989b). As discussed in Chapter 4, this programme is widely viewed as favouring road transport at the expense of other modes – particularly when it is noted that the majority of the capital is earmarked for the expansion of road capacity. Investment in rail services, on the other hand, is directed predominantly at the renovation of existing capacity (Potter, 1992b).

Roadbuilding schemes are justified using a cost-benefit analysis, taking into account congestion relief, reduced journey times, accident savings and environmental improvements. However, no comparable analysis is performed for alternative options, such as public transport investment. Rail investments have traditionally been evaluated using purely commercial criteria, based on annual rate of return. Understandably, the British government has received much criticism for not allowing a 'level playing field' on which road schemes can be evaluated against other possible solutions.

As part of a strategy for reducing CO_2 emissions, the traditional appraisal methods for road projects could be reviewed in order to take into account, among other things, global environmental implications. In particular, a 'CO_2 audit' might be performed for proposed road schemes, and compared with similar assessments for public transport alternatives. Such an assessment process for transport projects would need to be undertaken in a strategic context, as part of a target-led initiative for controlling greenhouse emissions.

Public transport

Evidence that investment in public transport can stimulate modal changes is scarce, although where such an investment has been made, the results are generally convincing. For example, a study of travel behaviour following the reopening of the Edinburgh to Bathgate line in Scotland in 1986 found that although 60 per cent of journeys undertaken on the new service were formerly undertaken by bus, around 23 per cent replaced car journeys (*Modern Railways*, 1986). Moreover, the modal change took place almost immediately after the reopening of the line, suggesting that most of the effect on travel patterns might be expected to take place in the short term, with further transfers taking place in the longer term. Further evidence for the modal change potential of public transport investment comes from an opinion poll by MORI which showed that in 1989, 32 per cent of drivers said that they would use their car less 'if public transport were better' (Corrado, 1990).

Investment in public transport can take a variety of forms, from major infrastructure projects such as new railway lines down to improvements in the quality of individual vehicles. In addition, it is widely recognized that investments in information technology can attract passengers to public transport. For example, it has been shown that bus passengers are concerned far more about knowing whether, and when, a bus will arrive than they are about the prospect of having to wait for one. Since 1992, bus stops on Route 18 in West London have been equipped with real-time electronic displays which provide passengers with information on how long they will need to wait for the next bus. The scheme, known as 'Countdown', is expected to spread to other bus routes in London, and similar systems are planned for other cities in Britain including Birmingham and Southampton.

Investment is not always sufficient on its own to achieve modal transfers away from cars to less energy-intensive forms of transport. There is a danger that, without policies to discourage car use, improvements in public bus and rail services could simply result in a larger number of trips being undertaken by existing users. Instead, investment policies should form part of a broad, target-led strategy for improving transport provision overall. For example, a scheme to develop light rail in Leeds has been designed as part of a wider strategy for switching commuter traffic from road to rail. It is intended that by 2010, every route into the city centre will have some form of rail alternative, with large park-and-ride stations on the city's outskirts.

Bus priority schemes

Transferring car trips to less energy-intensive modes may be brought about by a number of other investment programmes. Perhaps the simplest of all is the establishment of bus priority schemes. Figure 3.5 showed that buses are one of the most energy-efficient types of motorized travel available, and generally use less primary energy than all other motorized modes. In their basic form, bus priority schemes involve bus lanes from which car traffic is banned during peak hours. Other forms include bus-sensitive traffic signals, which can electronically detect the approach of a bus, and change to green in order to let the bus pass without stopping. At the moment, the progress of buses in many cities is obstructed by congestion, caused mostly by private cars. A key advantage of bus priority schemes in achieving modal transfer is that they can be implemented rapidly. In many cases, bus lanes already exist but require more effective policing to prevent abuse by car users.

Traffic calming

In the urban environment, there is considerable scope for reducing the environmental impacts of traffic through investment in traffic calming. In mainland Europe, traffic calming, or *Verkehrsberuhigung*, has for many years been a popular measure for improving the quality of residential and commercial areas (Whitelegg, 1990b). Traffic calming generally involves redesigning the layout of roads with the intention of reducing traffic speeds and making travel easier for pedestrians and cyclists. In many cases obstacles are introduced to car traffic, including speed humps, narrowings, gateways, chicanes and reduced speed limits. The movement of pedestrians is eased by the construction of road crossings and wider footpaths.[4]

The motives for introducing traffic calming tend to be related to road safety and the quality of the urban environment, and it is unlikely that traffic calming offers a significant potential for improving fuel economy. Chapter 3 showed that a typical car reaches its optimum fuel economy at a speed of around 70 kmh (44 mph), so measures to reduce the speed of urban traffic are likely to result in a small fuel consumption penalty.[5] However, the benefits of traffic calming almost certainly outweigh this minor penalty (Fergusson and Holman, 1990).

From the perspective of CO_2 emissions, the greatest value of traffic calming lies in the opportunities for modal transfer that it creates, both by deterring car use and by making conditions more suitable for pedestrians and cyclists. Traffic calming tends to affect short, local car trips most, and so the relative attractiveness of walking and cycling increases for these journeys. Short trips are also those in which the energy efficiency of the car is at its worst, and in which the energy-saving potential of modal transfer is therefore greatest.

The potential for modal transfers is high: in Britain, one in eight car journeys is less than 1.6 km in length (Department of Transport, 1988). Three-quarters of journeys are less than 8 km in length, yet most of these are undertaken by car. By investing in better facilities for cyclists, such as dedicated cycle lanes and secure parking areas, it should be possible to increase substantially the share of trips undertaken by bike. It has been estimated that if 20 per cent of non-walk trips were made by cycle, CO_2

emissions from transport could be cut by around 5 per cent (Rowell and Fergusson, 1991).

Traffic management

Chapter 3 described a system of traffic management that coordinates individual vehicles with the green phase of traffic lights, thereby reducing the number of occasions on which vehicles need to stop at junctions. Similarly, *route guidance* technology can allow drivers to navigate using electronic beacons or radio signals, obtaining information on the levels of congestion on different routes. By finding routes that are less congested, both average speed and fuel economy can be improved. A system known as TIGER (traffic information and guidance on European roads) is being developed by GEC, following the scrapping of its original 'Autoguide' system in 1992. Meanwhile, much of Britain's motoring network is now covered by 'TrafficMaster'. Developed by General Logistics, this system gives users real-time information on congestion, accidents and other delays, based on a network of infrared detectors mounted above the carriageway.

It is likely that there will be considerable demand from motorists for a device that would reduce their journey times in congested areas, and companies such as GEC are investing heavily in the necessary infrastructure. However, such systems offer little potential for reducing CO_2 emissions, given the high level of suppressed demand that exists in congested urban areas. Any extra capacity that is released is likely to be used up by additional vehicles entering the network (Mogridge, 1985).

Influencing demand

It is becoming widely accepted that in order to reduce congestion, it is necessary to tackle the *demand* for roadspace, rather than simply increasing the supply. A recent EC Green Paper on transport and the environment expressed the view that: 'If there is congestion, planning should not necessarily seek to increase the road network, because this can lead to an increase in demand' (European Commission, 1992).

Mention should also be made of the role of *marketing* in the promotion of public transport as an alternative to the car. In the 1980s, environmental concerns in Switzerland prompted a campaign designed to promote public transport as a more socially acceptable form of travel. Various market-based incentives were deployed, including positive advertising for public transport services, and special discounts and incentives for younger travellers. Similar campaigns were subsequently launched in Germany (Hass-Klau, 1990). But advertising is only successful if accompanied by a substantial level of investment in the transport system being promoted:

> ... a combination of investing in public transport and a different style of marketing will show results. But the public transport network has to be

very good, the interchanges excellent, the headways short and the facilities at high standards, all of which cannot be achieved without investing money.

(Ibid)

In summary, a wide range of measures is available for influencing both the demand for travel and the relative shares of modes that are used meet that demand. Given that technological measures alone are not sufficient to curb the projected rise in CO_2 emissions from personal travel, it is therefore possible to devise a strategy for controlling emissions which draws on both 'technical fixes' and measures to influence travel demand. In terms of the methodology introduced at the beginning of Chapter 5, this implies that policy measures from both Approach A and Approach B will be needed, introduced as a target-led package.

The IPCC has suggested that emissions of CO_2 from industrialized countries should be reduced by at least 60 per cent from the 1990 level in order to stabilize the atmospheric concentration of this gas. This target underpins the strategy developed below for reducing CO_2 emissions from personal travel. But before setting out policy measures to achieve this target, it is necessary to ascertain the reductions in emissions that should be undertaken by different sectors – transport, industry, domestic, commercial and so on. Indications from Germany suggest that, in the event of such an allocation, the transport sector would be allotted a less stringent reduction target than other sectors. It was suggested in Chapter 5 that if different targets were to be allocated to different sectors, transport might be required to cut its CO_2 emissions by 20 to 25 per cent as part of a strategy to stabilize the atmosphere.

The following sections evaluate the level of emissions that is likely to result from the introduction of an integrated strategy, embracing both technological measures and policies aimed at modifying travel demand.

Scenario 3: a strategy for sustainability

A third and final scenario has been devised for the SPACE model, designed to assess the scope for reducing emissions using a combination of 'technical fixes' and demand management policies. As before, the policies are examined in the context of Britain, but the conclusions drawn from the exercise are likely to be of relevance to all developed countries.

Different policy measures will have different geographical and sectoral impacts. Some, like area licensing, will have an effect predominantly in city centres, with little impact elsewhere. Others, like the redistribution of transport taxes, will affect the use of cars everywhere, although there may be geographical variations in response. Similarly, some measures will affect certain types of journey purposes, or groups of people, more than others. For example, traffic calming would tend to affect short, local car trips more than longer journeys.

Evidence from overseas suggests that policy measures for altering travel demand can be more effective, and more likely to be accepted, if deployed as part of a strategic package. For example, an area licensing scheme is more likely to be successful, and to be accepted by the public, if accompanied by an increase in the availability of high-quality public transport. Scenario 3 draws together such a collection of policy measures, none of which is regarded as being excessively radical in its own right. The philosophy underlying this scenario is that a 'package' of largely tried and tested policy measures is a promising approach to reducing CO_2 emissions, particularly when public and political acceptability are taken into account. By contrast, there is a danger that individual policies applied in severe form, such as a several-fold increase in fuel price, may be regarded as unacceptable both by governments and by the people that they are elected to serve.

Significantly, the British government has recently given its support to the principle of 'package' funding, whereby local authorities are able to bid to central government for a package of transport schemes, rather than financing individual projects in isolation. The move appears to indicate a shift towards a more integrated approach to transport planning.

An example of a comprehensive collection of measures aimed at reducing urban congestion is the scheme introduced in Singapore in 1975, which comprised:

- Area licensing for the central urban zone
- An increase in the number of buses in operation, and dedicated bus lanes
- A 67 km rapid transit system with 42 stations
- A new bypass avoiding the city centre
- Traffic management measures including 'smart' traffic signals, parking regulations and one-way systems
- Substantial taxes and restrictions on car ownership.

Although not all these measures would be regarded as politically acceptable in most OECD countries, particularly the rationing of car ownership, it is clear that this package has provided a solution to many of Singapore's traffic problems. The net result has been a substantial reduction in the volume of traffic entering the city during the morning peak period and exiting in the evening. Traffic speeds have been increased by as much as 30 per cent, while the modal share of travel held by car has been halved. Between 1974 and 1990, car traffic was reduced by 65 per cent (MTRU, 1991). Singapore exemplifies the complementary, synergistic effect of packaged policy measures, particularly in the urban context.

Assessment of policy measures

The following section describes the policy measures that have been adopted in Scenario 3 in addition to the technological changes introduced in Scenario 2, and assesses them in terms of cost, acceptability and timescale. The aim of this final scenario is to assess the potential for reducing emissions by all means available, namely technological measures, modal transfers, and reductions in travel demand.

This analysis is not intended to provide exact predictions of what will happen in Britain as a result of each policy measure; instead, it takes the form of an evaluation of the likely effects based largely on empirical evidence from schemes already in operation. The purpose is not to develop each policy area in fine detail, but to identify which measures or combinations of measures have the greatest potential to contribute to a strategy for reducing CO_2 emissions.

Redistribution of transport taxes and a carbon tax

This chapter has proposed, among other things, that the present system of transport taxation be rearranged in order to encourage energy conservation. One of the basic principles is the replacement of 'flat-rate' taxes with charges that reflect better the environmental impacts of personal mobility. Also recommended is the introduction of a carbon tax on fuels, which would serve to encourage the use of less carbon-intensive sources of energy. Scenario 2 showed how environmental taxation could be used to affect the choice of vehicle, in the form of incentives for consumers to purchase economical and alternative-fuelled vehicles. The principle is extended in Scenario 3 and used to influence not only the characteristics of the vehicle stock but also the nature of travel demand.

It is assumed that Vehicle Excise Duty (VED), presently charged at £125 per year for cars, is eliminated altogether and replaced with an additional fuel tax. This would make fuel economy a more important aspect of car purchasing decisions than at present, and so reduce the average fuel consumption of cars entering the market. This measure would also encourage more economical driving practices, as well as other effects such as reduced journey lengths and transfers to other modes. The VED charged on cars currently raises approximately £2.5 billion per year. At current rates of consumption, this amount could be raised by adding approximately 8p to the price of a litre of fuel, or 36p per gallon.

A further addition to the price of transport fuels is brought about in Scenario 3 through the introduction of a carbon tax. One of the benefits of this tax is that it affects a broad variety of activities, sending a general but clear energy conservation message to consumers. The travelling public is presented with a variety of options for reducing fossil fuel consumption, including using a more economical car, trip sharing, transferring to less polluting modes of transport, or reducing their overall travel distance. Some people may make no changes to their travel behaviour at all, and instead pay the additional cost.

The EC has proposed the gradual introduction of an energy tax equivalent to $10 on a barrel of oil, to be introduced progressively over the period 1993 to 2000. This approximates to 7p added to the price of a litre of petrol, and would generate revenue equally from non-renewable fuels and from fossil fuels according to their carbon content (Gardner, 1991). It is proposed in this scenario that the EC energy tax is introduced in the form currently proposed.

The combined effect of transferring VED to fuel taxes and introducing a new carbon tax would therefore be to add approximately 15p to the price of a litre of petrol (68p per gallon). This represents a fairly modest price increase, and it should be noted that a much larger increase in fuel price, of the order of 500 per cent rather than approximately 30 per cent as proposed here, would have a vastly more significant influence on travel behaviour. However, the approach underlying this modelling exercise is to rely on realistically modest measures which, when packaged together, are likely to be of value in curbing CO_2 emissions.

The increase in fuel prices is translated by the SPACE model into a reduction in car travel, based on an estimate of the elasticity with which car travel responds to changes in fuel price (Goodwin, 1988). This reduction is distributed four ways between transfers to other modes, car sharing, reduced journey lengths and eliminated journeys. In doing so, the SPACE model also takes into account the relative elasticities of different journey purposes: 'work', 'shopping and personal business', 'social and entertainment', and 'holiday and other'. For journeys to work, it is expected that the routine nature of the travel, coupled with the relatively easy availability of public transport, would lead to a relatively high elasticity of traffic with respect to fuel price. However, two factors are likely to act against this. Firstly, the opportunity for modal transfer may be limited as a result of capacity constraints in the public transport system.[6] Secondly, although there may be opportunities for modal transfer, there is unlikely to be a significant potential for trip shortening or elimination, as commuters tend to be 'tied' to particular workplaces and working hours. It is therefore expected that work trips will be less elastic than shopping and personal business trips.

In contrast with workplaces, most shops or services can be found at a choice of locations, so shopping trips are likely to respond more strongly to increased fuel price. Meanwhile entertainments and social trips, as well as holiday and 'other' journeys, are thought to lie somewhere between the two in terms of their elasticity.

It is also expected that the elasticity of car traffic will vary between different types of area. For people living in rural areas, the opportunities for reducing car travel are expected to be fewer than in urban and suburban areas.

For Scenario 3, it is assumed that the increase in fuel price would lead to a 'freeze' on car occupancy levels at their 1990 values, bringing to a halt the gradual decline in load factors that has taken place in recent years. The increased fuel cost is assumed to encourage a degree of car sharing for all trip types, stabilizing car occupancy from the year 1995 onwards.

The remaining reduction in car traffic is assumed to be distributed equally between modal transfer, journey shortening, and journey elimination.

An increase in fuel price would also have an effect on the average fuel economy of the car stock, according to the elasticity of fuel economy with respect to fuel price. In Scenario 3, therefore, car fuel economy increases more rapidly than in either of the previous two scenarios.

The important issue of equity should be addressed when considering increases in the cost of motor fuels. A common argument against raising the price of transport fuels is that people who are highly dependent on a car, such as those living in rural areas on low incomes, will suffer unduly as a result of any increase in price.[7] However, the package of measures proposed here offers most people benefits as well as costs, including an improved standard of public transport in all areas. If necessary, it is also possible for rebates to be made available to those groups in society for whom changes in the cost of travel are regarded as excessively harsh.

Enforcement of speed limits

This chapter has proposed a policy of fully enforcing the national 70 mph (113 kmh) speed limit in order to improve vehicle fuel economy. The effect on fuel consumption of a vigorous enforcement policy would be immediate. It is estimated that effective enforcement, using both police patrols and automatic surveillance, would improve average car fuel economy by 2 per cent over a period of five years. The effect on motorcycle fuel consumption figures is assumed to be negligible, since motorcycles account for less than 1 per cent of CO_2 emissions from personal travel. Similarly, the figures for coaches are assumed to be unaffected by this policy measure, since most coaches are now fitted with speed governors.

The effect that better enforcement of speed limits would have on modal choice is unclear. On the one hand, a reduction in car speeds could make other modes more attractive. On the other hand, the enforcement of speed limits may actually increase traffic flows by smoothing out bottlenecks. Without doubt, the benefit of speeding in terms of reducing journey times is generally overestimated by drivers.[8] For Scenario 3, it is therefore assumed that the modal distribution of travel would be unchanged by the full enforcement of 70 mph speed limits.

An 'urban access package'

As illustrated by experience from Singapore, the success of traffic restraint policies can be maximized if they are introduced as part of a synergistic 'package' of measures. In Scenario 3, an 'urban access package' is adopted, consisting of five measures:

- Area licensing

- Increased spending on public transport
- Employer-based travel schemes
- Business taxes to reflect accessibility
- Urban planning to reduce travel demand.

The package of measures is expected to have most influence on work-related journeys, since work trips account for the majority of city traffic, particularly in peak periods.

Area licensing

Evidence from Singapore and Milan provides useful guidance as to the likely effect of area licensing schemes in Britain's major cities, but it should be stressed again that an increase in the capacity of public transport is likely to be a prerequisite for the success of area licensing schemes. Licensing may also be supplemented by traffic calming schemes and parking restrictions inside the designated zone.

Currently 20 per cent of Britain's working population in urban areas commutes into the centre of large towns and cities whose size could justify area licensing (OPCS, 1984). In rural areas, the figure falls to approximately 5 per cent. These figures can be taken to represent the fraction of work journeys into areas that have a potential for licensing. For shopping, personal business, social and entertainment trips, it is assumed that fewer trips would be affected, since these types of journey are easier to divert to alternative locations outside the designated zone. Trips in the category 'holiday and other' are assumed to be unaffected by area licensing, because they tend to be longer and less scheduled.

On the basis of evidence from Milan and Singapore, it is assumed in Scenario 3 that 40 per cent of car trips affected by an area licensing scheme would be eliminated. Some of these would be avoided altogether, while others would be transferred to bus, rail and cycle. Walking is not considered to be a significant alternative for the type of car trips that would be subject to area licensing. It is also expected that an area licensing programme would lead to an increase in car sharing, and a resulting rise in occupancy.

Public transport spending

The second policy measure in the urban access package involves an increase in investment and revenue subsidy for public transport. The essence of such a policy is to facilitate a transfer from car to public transport, which would also be stimulated by other elements in the package. However, transfers to public transport would also take place from walking and cycling, as a result of the reduction in fares.

The use of public transport responds positively to reductions in fares (Goodwin, 1988). The increase in public transport use that would take place as a result of fare subsidies in Scenario 3 would arise as a result of both transfers from other modes and the generation of additional trips. It should be noted that there is generally a degree of 'suppressed demand' on overcrowded urban transport systems, many of which are

operating at full capacity and beyond. A programme of capital investment would facilitate an expansion in capacity, thereby allowing an
increased level of patronage. An investment in extra capacity would
therefore result in increased patronage even without the additional use
of fare subsidies.

Clearly a reduction in public transport fares would have the effect of
transferring some car travel to bus and rail. However, it is also likely that
some new trips would be generated, and there would also be some modal
transfers from cycle and walking to public transport, as a result of the
lower fares. The situation in terms of CO_2 emissions is not, therefore,
straightforward, and a careful balance needs to be struck between
promoting modal transfers from cars, and generating additional trips. It
is likely that policies would need to be 'fine-tuned' at regular intervals, on
the basis of observed changes in travel demand.

Combining the area licensing policy with the increase in fare subsidies,
Scenario 3 assumes that the 40 per cent of eliminated car trips would be
distributed equally between new bus trips and new rail trips. Cycling is
not expected to increase as a result of area licensing, because the
simultaneous reduction in public transport fares would make bus and
rail more attractive to cyclists. Fares would need to fall by approximately
30 to 40 per cent in order to facilitate this transfer to public transport,
and there would need to be an expansion of public transport capacity in
order to accommodate the transfer to bus and rail.

Employer-based travel schemes

The third measure in the urban access package is a programme of
employer-based travel schemes, aimed at reducing the use of cars. As
described earlier in the chapter, measures pioneered in the US include
car sharing, bus provision and telecommuting. A policy aimed at
establishing similar schemes in Britain would be a long-term undertaking, although some results would begin to be seen fairly quickly.
Incentives for car pooling could reduce car mileage relatively rapidly,
whereas measures aimed at providing mass transit and promoting
teleworking would take longer to implement. One of the hazards of a car-
pooling policy is that it may simply attract public transport users into
cars, which in most cases would have a negative rather than a positive
influence on energy conservation. However, this danger could be
minimized by undertaking improvements in public transport alongside
car-pooling schemes.

There is little experience of car-pooling schemes in Britain, though a
scheme tested by the Automobile Association is regarded as having
largely failed in its objectives as a result of a lack of interest on the part of
the workforce. However, a number of demonstration schemes have been
successfully established in the US. A summary of pilot projects initiated
by employers in California is provided by Collier (1991), and the findings
are summarized in Table 7.1 above. It should be noted that these results
were achieved in a climate of very cheap fuel, and in a highly car-

dependent society. Conditions in most other developed countries, where fuel is cheaper and residential land use less dispersed, are likely to be considerably more conducive to modal transfer than those in California. In Scenario 3, 5 per cent of work journeys by car are eliminated after 5 years as a result of car-pooling measures, rising to 10 per cent of journeys after 15 years.

Between 2010 and 2025, an additional 5 per cent of work trips are eliminated as telecommuting becomes available for some employees. Meanwhile car occupancy increases for people living in urban and semi-urban areas, by 5 per cent over the period up to 2000.

Business taxes
The fourth component of the urban access package is the use of variable business taxes to reflect the relative accessibility of different locations. In the absence of empirical evidence, the effect of such taxes is difficult to predict, though in qualitative terms the likely result would be a transfer of business premises closer to residential areas and urban centres, and a reduction in the length of work journeys. In addition, it is probable that some shopping and personal business trips would also be shortened, as customers are able to gain access to these services closer to home.

For a particular town, a 20 per cent reduction in the average length of commuting trips appears to be an achievable target for such a policy as part of the urban access package. The average shopping or personal business trip is assumed to be shortened by only half this amount, since fewer journeys would be affected. Such a reduction in journey length would also allow some car trips to be transferred to bus, rail, walking and cycling. The timescale for full implementation is assumed to be 30 years, since relocation of existing businesses (and turnover of staff) cannot be expected to take place other than as part of long-term business relocation decisions.

Urban planning to reduce travel demand
The fifth and final element of the urban access package involves the implementation of planning regulations aimed at reducing travel demand. This policy would affect the location of homes, businesses, shops and services, as well as introducing new schemes such as traffic calming, pedestrianization and revised carparking arrangements. Of all the policies, this has the longest timescale. It is not realistic to suppose that changes in land-use planning could achieve their maximum potential within the timescale of the SPACE model, which extends to 2025. Nevertheless, there will be a substantial number of new developments taking place in this period, many of which could be located so as to prioritize access rather than mobility.

This chapter has suggested two types of urban form that might reduce the need for travel, based on work carried out by Steadman and Barrett (1991). The first is a concentrated, centralized city structure, and the second is an arrangement of small local centres or 'villages' surrounding a traditional urban centre. A move towards these types of layout could be achieved by central government working through the local planning system. To a large extent, a transfer to walking and cycling would be expected to follow naturally from any reduction in land-use dispersal, though it could be further encouraged by improvements in the facilities for these modes. At present, safety is a major deterrent to cycling in Britain, and it is likely that improvements in junction design, together with the creation of dedicated cycle lanes, could encourage a significant number of car users to transfer to non-motorized travel for local trips.

The benefit of such a policy would be a very gradual one, and would tend to affect journey lengths rather than the number of trips. There would also be a concomitant transfer away from cars to public and non-motorized travel modes, which itself would tend to reduce trip lengths. In Scenario 3, the effect is expected to take place across all types of journey, with the exception of holidays and other long-distance trips.

This policy measure is complementary to the four other components of the urban access package, and no attempt is made here to evaluate its effect in isolation. Rather, land-use planning is viewed as a backdrop against which other measures, such as area licensing and public transport investment, are more likely to succeed. Equally important is the long-term benefit of less travel-intensive settlement patterns, which will help to ensure that reductions in CO_2 emissions from personal travel can be sustained beyond 2025 and throughout the 21st century.

The environmental assessment of roadbuilding schemes

Up to this point it has been fairly straightforward to estimate the effect of the various policy measures deployed in Scenario 3, subject to a degree of uncertainty. The remaining policy measure, which involves a review of the method used to assess inter-urban road and rail schemes, is more difficult to evaluate. The measures contained in the urban access package are all concerned with travel in urban areas, though they will also have some influence on inter-urban trips. The final policy measure in Scenario 3 concerns inter-urban travel, and in particular the way in which different modes are considered by central government during the transport planning process.

Recent traffic growth in Britain has taken place mainly on inter-urban routes. In the decade to 1991, traffic on British motorways and other major rural roads grew by 78 per cent, whilst that on built-up major routes grew by a more modest 22 per cent (Department of Transport, 1992c). Although the public perception of growing traffic problems tends to focus on towns and cities, where the effects of congestion and air pollution are most noticeable, a major growth in 'business as usual' CO_2 emissions is likely to take place outside urban areas.

The government's trunk roads programme involves an expansion of the capacity of Britain's road network, based on a substantial forecast growth in road traffic (see Chapter 4). Scenario 1 has shown that the increase in CO_2 emissions resulting from this 'business as usual' growth is almost certainly incompatible with a national strategy for reducing emissions. Nevertheless, a limited growth in traffic volume may be possible as part of a CO_2 reduction strategy.

The Department of Transport's White Paper *Roads for Prosperity* (Department of Transport, 1989b) outlines the government's commitment to roadbuilding, and gives details of an expanded trunk road programme for Britain. The paper states that '...the main way to deal with growing and forecast inter-urban road congestion is by widening existing roads and building new roads in a greatly expanded road programme'. However, building roads also tends to increase the demand for roadspace, and in some cases leads to worse congestion in the urban areas into which they feed traffic. This was recently expressed by the European Commission in its Green Paper on transport and the environment, which argued: 'If there is congestion, planning should not necessarily seek to increase the road network, because this can lead to an increase in demand' (European Commission, 1992).

It is reasonable to expect that a reduced budget for roadbuilding would result in a more modest growth in road traffic than that currently forecast. Coupled with an increase in government spending on public and non-motorized modes, a reduction in road spending could also stimulate modal transfer to rail and inter-urban coach travel. How the growth in car traffic would be constrained would depend on what other policies were adopted. In the absence of other policies, traffic levels would be held down by increasing levels of congestion. Alternatively, an investment in public transport could facilitate a transfer from the car to other modes of travel.

The final component of Scenario 3 is the establishment of a common appraisal system for road and rail schemes, based on a full cost-benefit analysis including consideration of relative CO_2 emissions. Such a policy has been advocated by the Royal Town Planning Institute (1991), which suggests that: 'The application of consistent mechanisms of assessment in economic, social and environmental terms to all transport proposals, together with an opportunity for review by an independent body, is an objective that should command wide support.'

Transport consultants Steer, Davies and Gleave have recently assessed the feasibility of transferring capital from the proposed M1 motorway widening scheme to the electrification of British Rail's London to Sheffield main line. They estimated that patronage of the line could be doubled over the next 20 years, with 60 per cent of the new passengers switching from car. By using the cost-benefit analysis method currently associated with road appraisal, the consultants showed that the financial benefits of the scheme were ten times greater than if the conventional appraisal method for rail projects had been used (Black, 1992).

It is impossible to predict what proportion of road projects might be cancelled in favour of less carbon-intensive travel modes, and therefore the amount of capital that would be available for investment in other modes, if a common investment appraisal process were to be adopted. However, in the example of the M1 motorway above, the result of transferring capital from road to rail investment is a 6 per cent transfer of traffic from motorway to rail as a result of the improved rail service. If such a strategy were to take place at a national level, then the modal transfer would be likely to be greater than this because of 'network' effects and more deep-rooted changes in travel behaviour. The effect would be twofold: existing journeys would be transferred to rail from road, and the future growth in car traffic would be diminished. On the basis of the M1 example, Scenario 3 assumes that the introduction of a common investment appraisal process would yield a 20 per cent reduction in car traffic growth up to the year 2025, of which half would be transferred to rail. The reduction is expected to take place for all journey types. However, the modal transfer to rail is not expected to take place among people living in rural areas, because they would be less likely to have ready access to inter-city rail stations.

An expansion of inter-urban rail capacity, including the introduction of new high-speed lines, would also be likely to attract a modal transfer from air to rail. Accordingly, Scenario 3 assumes a 20 per cent transfer of passenger kilometres from domestic air services to high-speed rail, over a period of 20 years. However, a policy of expanding inter-urban high-speed rail would need to be monitored carefully, because airline companies might be tempted to run rail services as a substitute for flights as part of a policy to increase the number of long-haul flights that they were able to operate within the constraints of limited airport capacity. The net effect of such a policy would be detrimental in terms of CO_2 emissions.

Constructing a stabilization scenario

Having evaluated all the policy measures available for restraining the demand for travel, it is possible to use the SPACE model to assess the combined effect of the 'technical fixes' set out in Chapter 6 and the demand restraint measures described in this chapter. Scenario 3 is designed to test the effect of pulling out all the stops, in terms of realistic, tried-and-tested policy measures.

The overall target for curbing CO_2 emissions, as proposed by the IPCC, involves a reduction of least 60 per cent in order to stabilize the atmospheric concentration of this gas. It was suggested earlier that transport might be allocated a target of reducing its CO_2 emissions by 20 to 25 per cent as part of this 'atmospheric stabilization' strategy. What reduction in CO_2 emissions does Scenario 3 deliver?

Figure 7.3 shows the effect of Scenario 3. Total CO_2 emissions from personal travel are reduced by 22 per cent, a figure which corresponds remarkably well with the objective identified above. Significantly, the majority of the reduction takes place up to the year 2000.

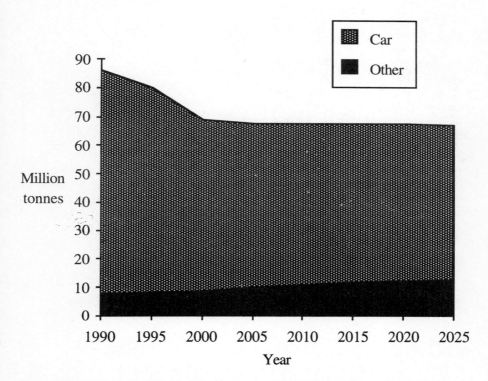

Figure 7.3 Total emissions of CO_2 from personal travel in Scenario 3

This result suggests that Scenario 3 offers a means by which personal travel may contribute to a strategy for stabilizing the atmospheric concentration of CO_2. In other words, if other energy sectors besides personal travel were to play their part in a CO_2 reduction strategy, then the atmospheric concentration of this gas could realistically be stabilized.

It is particularly important to note that the measures contained in Scenario 3 have all been conservative in magnitude. Estimates of the effectiveness of policies have been realistic, and have not involved decimating the car population or rationing personal mobility. On the contrary, the substantial cut in emissions has been brought about by the complementary, synergistic effect of various measures, most of which have been successfully demonstrated in some form throughout Europe, Asia and North America. Intriguingly, the 22 per cent reduction in CO_2 emissions is achieved at the same time as a modest growth in car traffic – 16 per cent over the period 1990 to 2025 – as Figure 7.4 illustrates.

In short, Scenario 3 offers a realistic package of measures for reducing travel-related CO_2 emissions to a level that is compatible with atmospheric stabilization.

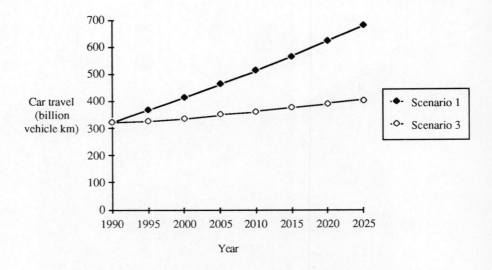

Figure 7.4 The effect on car traffic growth of Scenario 3 measures

Notes

1. Policy measures for reducing energy demand do not, however, necessarily depend on increased densities. A variety of structures are available that combine relatively low densities with high levels of public and non-motorized travel. Medium-density settlements of individual houses with gardens can be compatible with reduced car use, if combined with careful land-use planning designed to reduce travel need.
2. The highest travel intensity among individuals in England and Wales occurs in a large ring in south-east England, centred on London (Banister and Banister, 1991).
3. The company also estimates that a trans-Atlantic trip by air could be replaced by five weeks on the telephone for the equivalent energy cost!
4. In March 1992, the Traffic Calming Bill in Britain received Royal Assent, providing a legal framework in which local authorities may install traffic calming schemes. The result is likely to be a move away from the use of road humps, which are not always suitable for traffic conditions, towards other forms of obstacle such as chicanes, gates and shared spaces.
5. The nature of many traffic-calming schemes is such that cars are required to accelerate and decelerate frequently. Cruder designs, such as speed humps, may actually encourage harsh and uneconomical driving practices. A consideration of car fuel economy should ideally be taken into account in the design of traffic-calmed streets.
6. This effect can be seen on the rail network around London, which in some areas is used beyond capacity.

7. The British government came close to electoral disaster in 1993 when it proposed the introduction of Value Added Tax on domestic heating fuels. Such a policy, albeit intended to raise revenues, was viewed by some as a near-perfect example of environmental taxation, and yet public hostility to the plan was fierce.
8. A typical business trip is around between 50 and 100 km in length, for which a 15 kmh reduction in speed would lengthen the journey time by no more than five minutes.

8

A 12 Point Plan for Sustainability

If you want someone to clean the yard, go out and clean it.
And if they agree with you, they will come and help you.

Chief Oren Lyons

The conclusions to be drawn from the previous chapter are encouraging. If it is assumed that personal travel should be allocated a less stringent target than other sectors for reducing CO_2 emissions – as in the case of the German programme – then it is possible for this sector of the economy to achieve what might be regarded as a 'sustainable' level of greenhouse gas emissions. This can only be done, however, through a committed strategy that combines technological changes with a package of demand management policies. The 12 policy measures that collectively comprise Scenario 3 are summarised in the box overleaf. Table 8.1 identifies the main areas in which each policy is expected to exert an influence.

Timescales of policy measures

The 12 policy measures put forward in this strategy have a wide range of timescales, or 'lead times'. Technological measures that can be applied to vehicles are constrained principally by the rate of turnover in the vehicle market. Typically a car has a lifetime of between 10 and 15 years before it is replaced. More economical vehicles and alternative fuels could therefore only be introduced gradually, with a time-lag of approximately 10 years between the introduction of a policy and its ultimate fruition. But technological measures can also be delayed by institutional processes. Regulations and major changes to the taxation system, such as the introduction of a carbon tax, can take several years to be approved by national government and set in place. By contrast, adjustments to the existing taxation structure, such as rebates and surcharges related to fuel economy, could be introduced almost immediately in the annual budget statement.

---

12 point plan for 'sustainable' emissions from personal travel

1. Introduction of fuel economy labelling for new cars, and regulations on the content of car advertisements
2. Introduction of a 'feebate' scheme for new cars based on fuel economy
3. Establishment of a long-term programme for the research, development and deployment of 'greenhouse-neutral' transport fuels, including incentives for consumers
4. Enforcement of road speed limits, and possibly downgrading some existing limits, to improve average fuel economy
5. Introduction of regulations requiring employers to reduce the travel demand of their workforce
6. Establishment of an environmental assessment programme for proposed road schemes, whereby CO_2 emissions are compared under a number of different options
7. Increased investment in, and subsidy for, public and non-motorized modes of transport (partly funded by cancelled roadbuilding schemes)
8. Redistribution of transport taxes to promote fuel economy, alternative fuels, modal transfer and reduced travel volume
9. Business taxes to be related to a measure of accessibility for customers and employees
10. Introduction of area licensing to discourage the use of cars in urban areas
11. Establishment of strategic planning regulations for promoting local access, rather than encouraging highly mobile and energy-intensive lifestyles
12. Introduction of a carbon tax to reflect the environmental cost of burning fossil fuels, and to reduce consumption

---

Other policy measures for reducing CO_2 emissions are, by their very nature, more long term in their influence. In particular, a policy of travel reduction through land-use planning would take many years to achieve results, because of the delays involved. Even employer-based schemes for reducing travel demand would take time to have an effect, since most employees are unlikely to alter their lifestyles and travel arrangements instantaneously. More immediate impacts could be achieved through measures such as bus priority schemes.

Although the more radical measures such as land-use policy tend to have longer lead-times, this does not imply that they should be given less priority, or adopted with any less urgency, than shorter-term measures. On the contrary, an effective strategy might involve prioritizing the longer-term measures, at the same time as instituting more immediate policies in the shorter term. In other words, short-term policies such as incentives for fuel efficiency can be used to 'buy time' while more long-term measures such as land-use planning are set in motion. All this indicates the need for a strategic, target-led approach to reducing transport emissions, rather than a collection of piecemeal, short-term measures. The synergy between different policy measures is often as important as the measures themselves.

Table 8.1 Areas of influence for each of the policy measures

Policy measure	Areas of influence			
	Fuel economy	Alternative fuels	Modal transfer	Reducing travel demand
1. Fuel economy labelling/ advertising	•			
2. Fuel economy 'feebates'	•			
3. Alternative fuels programme		•		
4. Speed limits	•		•	•
5. Employer-based travel schemes	•		•	•
6. Environmental assessment of transport projects			•	
7. Increased spending on public transport			•	
8. Redistribution of transport taxes	•	•	•	•
9. Business taxes to reflect accessibility			•	•
10. Area licensing and road fees			•	•
11. Land-use planning for reduced mobility			•	•
12. Carbon tax	•	•	•	•

Public acceptance of policy measures

Public acceptability is a vital requirement of any strategy for reducing CO_2 emissions from personal travel, for both ethical and practical reasons. It would be unjustifiable – and politically foolhardy – to go about implementing a CO_2 abatement strategy for which there was no public support. There is evidence that governments have fought shy of tackling the environmental impacts of mass mobility, fearing a fierce backlash of public opinion, and instead opted for the 'technical fixes' approach. In the words of Cragg (1992):

> Since the public are reluctant to give up the car and the flexibility of mass-transit systems is limited while the cost is high, governments – rightly or

wrongly – regard the reduction of pollution from motor vehicles as a research and development problem for the motor manufacturers and the oil companies.

This does not mean, however, that we are doomed to a 'business as usual' global greenhouse. If policies aimed at restraining mass mobility prove unpopular with the voting public, it will be as much to do with the way in which the policies were presented as it will do with the policies themselves. Countless opinion polls have borne out this observation. Public reaction to a particular policy can range from warm enthusiasm to outright hostility, depending on the way in which the proposal is presented.

For example, traffic calming schemes have been received much more readily in instances where the people affected by them are first consulted on the proposals. For obvious reasons, very few people would be happy to wake up in the morning and find the physical structure of their street being redesigned, with no prior discussion. On the other hand, traffic calming schemes instituted with the involvement of local residents generally prove highly popular. Similarly, urban road pricing is demonstrably far more popular if the public knows that the revenue raised will be used to improve local bus services, rather than disappearing into the black hole of central government. When a sample of Londoners were asked whether they would support road pricing in the capital, 43 per cent of respondents supported the notion, with 53 per cent against it. But when it was suggested that the money raised would be used to reduce congestion, the proportion in favour rose to 62 per cent (MacKinnon, 1991).

The context in which policies are introduced thus exerts a critical influence on public acceptance. These observations support the view that policies should be deployed as part of a wider strategy with a clearly defined purpose. By making it clear to the public what the overall aim of the strategy is, and what the benefits are expected to be, it is possible to create a more favourable impression of individual policy measures.

The 12 point plan set out in this chapter should not be regarded as a list of policies from which governments might pick and choose according to preference. There are many interactions between the different policy measures, whose net result is to make the whole package considerably greater than the sum of its parts. All the evidence suggests that individual policy measures are more likely to succeed if deployed as part of a 'package' of measures. The packaging of policies has two benefits: firstly, the synergies that exist between various policies can be exploited, and the overall effect enhanced; and secondly, public acceptance of policy measures is more likely to be won if the measures form part of a strategy with a clear purpose. In addition, the varying timescales of different policy measures can be given full attention if the policies are assembled into a single package.

9

Conclusions

We are at the very beginning of time for the human race. It is
not unreasonable that we grapple with problems. But there
are tens of thousands of years in the future. Our responsibility
is to do what we can, learn what we can, improve the solutions,
and pass them on. It is our responsibility to leave the people of
the future a free hand.

Richard Feynman

The work described in this book was prompted by an awareness that
personal travel is a significant, and growing, contributor to emissions of
'greenhouse' gases, particularly CO_2. Current forecasts throughout the
OECD point to a continuing growth in the amount of travel undertaken
by each person, and a steady increase in energy consumption and CO_2
emissions. Given that the historical growth in personal travel has been
mainly the result of longer journeys, rather than additional journeys,
there is an urgent need to examine what benefit, if any, this projected
growth in mobility will actually deliver.

Under the umbrella of the Climate Change Convention, national
governments world wide have formally acknowledged the reality of the
global warming threat and the need to take precautionary action to avert
a climate catastrophe. Although the level of response so far agreed by its
signatories is modest, the Convention provides a vital framework under
which future targets can be negotiated and agreed. Climate change is a
global problem which requires a global response. It is hoped that the
conclusions of this book will help to establish policies aimed at combining
the provision of personal transport with the protection of the global
environment.

Principal findings

Scientific opinion strongly favours the hypothesis that human activities are beginning to change the climate through emissions of so-called greenhouse gases. Although the evidence is currently inconclusive, and likely to remain so, climate measurements are broadly consistent with the view that significant warming has begun. The need to find ways of curbing greenhouse emissions is likely to become greater, not less, as the warming trend emerges. The most comprehensive survey of the science of global warming is that undertaken by the Intergovernmental Panel on Climate Change (IPCC) scientific working group, which estimates that an immediate reduction of at least 60 per cent in CO_2 emissions will be required in order to stabilize the atmospheric concentration of this gas and to avert the threat of global warming.

Personal travel is a major contributor to CO_2 emissions in Britain, as in the rest of the OECD, and its significance will increase as the demand for travel continues to rise. A 'business as usual' scenario, as examined in Chapter 4, implies a near-doubling of CO_2 emissions from the transport sector. To date, the British government has not put forward any means by which such an increase in CO_2 production might be reconciled with its target for stabilizing emissions overall.

In the field of personal transport, there are many options available for breaking the upward trend in CO_2 emissions. These may be broadly categorized into two types of policy measure: technological changes and modifications in the demand for travel. A technology-based strategy, involving a combination of fuel economy incentives and alternative fuels, would lead to a reduction in the rate at which emissions rise. But the total output of CO_2 in Britain under a 'technical fixes' scenario would still be 35 per cent higher in 2025 than it was in 1990.

It is therefore necessary to look beyond technological changes in order to achieve a reduction in CO_2 emissions from personal travel. A variety of policies are available for influencing the volume and modal distribution of travel. Some of these are measures that have been used in the past for other purposes, such as the subsidization of public transport. Others are unprecedented, such as the European Commission's proposed carbon tax.

Reducing the demand for travel does not automatically imply the use of 'trip suppression' measures, but rather the promotion of lifestyles that do not require the level of mobility upon which society is presently dependent. Like energy, transport is rarely something that people value in itself; more commonly, it is a means of gaining access to people, goods and services.

By combining a number of policy measures together into a third and final scenario, it has been possible to estimate the combined effect on CO_2 emissions of technological changes plus policies to moderate the demand for travel. The modelling experiment indicates that such a strategy could reduce CO_2 emissions from travel by 20 to 25 per cent. A number of the policies contained in this scenario are deployed in the form of a 'package', in order to exploit their complementary effect on one another.

Implications for an overall CO_2 reduction strategy

Some of the elements contained in the final scenario are, to some degree, speculative, and the 20 to 25 per cent reduction in emissions that the SPACE model predicts should not be regarded as a representation of the greatest reduction in emissions that is possible. Rather, it is the best estimate of the aggregate effect of the 12 policy measures proposed in the final scenario. Without the benefit of real-life experience, there is inevitably a degree of uncertainty surrounding the true potential for reducing greenhouse emissions.

If it is assumed that personal travel should be allocated a target less stringent than those for other energy sectors, it appears possible for Britain to achieve a 'sustainable' reduction in CO_2 emissions across all sectors. The question of allocating sectoral targets is difficult and requires further research, but some work has been undertaken to this end in Germany. As part of an overall 25 per cent reduction in CO_2 emissions, the transport sector was apportioned a target of 9 per cent. If the same ratio were applied to the 22 per cent reduction in emissions projected for personal travel in Scenario 3, it would imply an *overall* reduction in emissions of 61 per cent. This corresponds almost exactly to the IPCC 'stabilization' target of 60 per cent overall.

The conclusion to be drawn from these speculative calculations is that the maximum possible reduction in CO_2 emissions identified in this study could be consistent with the IPCC target of 'atmospheric stabilization'.

The Netherlands National Transport Plan

Many of the principles put forward in this book are echoed in the practical policy measures proposed under the Netherlands National Transport Plan, which has been formally adopted by the Dutch government. The *Tweed Structuurschema Verkeer en Vervoer*, or SVV2+, differs from previous transport plans in that it includes measures to curb the projected growth in car travel. The issue of global warming is addressed by a national target to reduce transport-related CO_2 emissions from 23 to 21 million tonnes by the year 2010 (Sturt, 1992).

A 'business as usual' scenario for car travel in the Netherlands, as presented in SVV2+, anticipates a 70 per cent growth in car kilometres between 1986 and 2010. The document states that this hypothetical growth will, in reality, be constrained by the national pollution targets, as well as financial and spatial limitations. As a result, it is suggested that the overall growth up to 2010 be halved to 35 per cent. By comparison,

Scenario 3 in this study envisages a 14 per cent growth in car traffic between 1988 and 2010 (see Figure 7.4).

There are clear parallels between the Dutch approach and the ideas developed in this book. Technological measures are, in both cases, viewed as inadequate to achieve CO_2 abatement targets, and policies are sought to constrain the projected increase in travel demand.

To this end, measures proposed by the Dutch plan include sharp increases in fuel taxes and major investment in public transport infrastructure. Plans to expand the Dutch rail network are already under way. Similarly, the final scenario of this study proposes a large-scale expansion of rail capacity for inter-urban trips, in order to achieve a substantial modal shift away from car travel. The Dutch plan also draws heavily on land-use policies to help limit the growth in car travel, as described earlier in the book. Future employment centres would be classified according to the nature of their transport links, and different types of activity assigned to appropriate areas.

The Dutch programme, although differing in fine detail from the strategy developed in Scenario 3, has much in common with it. It demonstrates that many of the policy measures identified by this study are entirely feasible as part of a strategy for curbing motorized travel. More importantly, the lessons learned from the Netherlands can be used to help shape similar strategies throughout Europe, Australasia and North America. It is possible that some elements of the Dutch plan will be more successful than others; but by pushing ahead with its environmental transport strategy, the Netherlands has provided a vital focus for the further development of policies aimed at curbing CO_2 emissions.

Non-operational emissions of CO_2

Substantial quantities of CO_2 and other greenhouse gases are produced not in the operation of transport, but in peripheral activities such as manufacturing and maintaining vehicles and infrastructure. This study has given some attention to the emissions produced by these secondary sources, but has not included them in the three scenario projections. This is because primary fuel consumption for transport operations represents a much larger use of energy. However, it is possible to make a number of observations about the future level of these secondary emissions.

Firstly, the scope for reducing emissions from these sources lies largely outside the transport sector, in the electricity supply industry. A large-scale transfer away from coal to less carbon-intensive sources of electricity would reduce substantially the level of CO_2 emissions from 'secondary' processes. Secondly, it appears unlikely that the demand for these processes could be curbed other than as a result of a reduction in travel demand. In particular, the energy and resources required for car production and road maintenance would be likely to decrease only if there were a reduction in the number of car kilometres travelled, and in the turnover of the car stock.

Beyond 2025

The three scenarios in this study all run to the year 2025, and take no account of trends that might prevail after that year. It is likely that reductions in CO_2 emissions could continue beyond 2025 if renewable energy were to increase its role in personal travel, and if land-use planning continued to embrace an access-led approach. However, such progress would depend critically on the groundwork done in the period leading up to 2025.

The ultimate potential of alternative fuels for eliminating travel-related greenhouse emissions depends heavily on national energy policy. A programme to develop renewable power sources, in the form of wind, wave, tidal and hydroelectric energy, would facilitate a large-scale transfer from petroleum to renewables in the transport sector. If global warming does not prompt a widespread move away from fossil fuels, it seems likely that supply shortages will eventually do so.

Both short-term and long-term policies need to be adopted as a matter of urgency. Longer-term measures require extended lead times, and short-term measures may be regarded as 'buying time' whilst long-range measures are implemented.

Which way now?

The majority of climatologists agree that the global warming threat is almost certain to become more severe in a 'do-nothing' world. Even if anthropogenic greenhouse emissions were to cease immediately, global temperature and sea level would continue to rise as a result of the greenhouse 'commitment' that has already been made. The efforts of the IPCC have provided a sound scientific foundation upon which to base future policies, and it is now the responsibility of all governments in the developed world to address the issue of climate change with a sense of vision and urgency. To date, the response of most governments to the issue of climate change has been characterized by strong words but little action, and it is hoped that the preliminary consensus achieved by the Climate Change Convention can be developed into more stringent targets for curbing greenhouse emissions.

At the time of writing, transport policy in Britain is undergoing major changes, particularly in the way that investment for different modes is assessed by central government. Street-running trams have returned to Manchester, and more are expected soon in Sheffield, with plans for a line in Birmingham well advanced. The SACTRA report (Department of Transport, 1992d) has outlined ways in which environmental protection may be built into the trunk road appraisal process, and the Departments of Transport and the Environment have begun to examine the role of land-use planning in reducing the need for motorized travel. In addition, the Department of Transport has agreed, for the first time, to consider 'package' bids from local highway authorities, whereby a portion of their annual Transport Supplementary Grant is typically awarded not to individual schemes but for region-wide integral transport strategies. This change in priorities also allows local authorities to collaborate on large-scale public transport projects.

To some extent the new emphasis on integrated transport planning is the result of a realization that sustained traffic growth is incompatible with environmental protection goals, as well as with the efficient running of towns and cities. The European Commission should take some of the credit for bringing environmental considerations into the national transport policies of member states. Its Green Paper on sustainable mobility (European Commission, 1992) cautions against using road-building as a means to ease traffic congestion.

Even in car-crazy southern California, substantial sums of money are being diverted away from roadbuilding and into the construction of new commuter rail networks and street-running tram services. However, it seems absurd that massive subsidies should be required in order to persuade the residents of Los Angeles to use their new railways when the greatest obstacle faced by public transport in North America is the staggeringly low price of gasoline, which makes car travel impossible to beat as a low-budget form of transport. Obvious though this may seem, the point appears to have been lost on many transport planners.

Technological change, in the form of more efficient cars and alternative sources of energy, offers the potential for substantial reductions in CO_2 emissions from personal travel. But the vision of clean, environment-friendly cars is based more on wishful thinking than on hard reality, and there is a danger that an over-emphasis on technological solutions will overshadow the need for more fundamental changes in the nature of travel. It is hardly surprising that the motor industry, acutely aware of the rising tide of environmental concern, has sought to promote solutions based on the concept of an environment-friendly car, rather than allowing the culture of the automobile to be undermined. However, the main issue at stake is not the fuel economy of next year's Ford, but the enormous growth in car ownership and personal mobility that has taken place over the last 40 years, and which is forecast to continue for the next 40.

The actual benefit offered by this mobility revolution is highly questionable, given that the growth is mainly the result of longer journeys that are made to undertake the same activities. A more strategic, less market-led approach to land-use planning offers tremendous potential for reducing daily travel, particularly by car, and releasing time that was previously spent behind the wheel. Few people would dispute the pleasure of living in a neighbourhood which offers all the necessary conveniences within walking distance of home.

Access – to people, goods, services and facilities – should be available to everyone, regardless of income, class or background. But the right to free access has often been mistakenly interpreted as a right to unlimited mobility – and the result, paradoxically, has been to polarize society into those with access to a car and those without. Unrestricted mobility is something which clearly cannot be made available to everyone within the existing constraints on space, resources and air quality. So while the better-off have been able to acquire more cars and travel further, the effect on those without access to a car has been to reduce their access, by contributing to the decline of public transport and the creation of sprawling settlements which are hostile to walking and cycling. When it is no longer possible to walk to the end of the street to buy a loaf of bread, and instead a car must be used to travel 5 miles to the nearest shopping centre, who is benefiting?

Although the car has brought undisputable benefits to a large number of people, its main influence – through pollution, casualties, nuisance and social division – has been to undermine modern society rather than civilise it. The 'new realism', based on the idea that roadbuilding is an inappropriate instrument for dealing with problems of congestion and pollution, contrasts strongly with the view traditionally held by the British government, which commonly cites environmental protection as a *justification* for building more roads. For example, many elements of the government's roadbuilding programme are presented as bypass projects, for which reductions in CO_2 emissions are claimed as one of the benefits. No reference is made to the likely traffic-generating effect of increased road capacity.

From the point of view of resources, one might question the wisdom of a transport policy that is based on the assumption of a continued growth in car ownership, personal mobility and energy consumption, given that oil reserves are already beginning to dwindle as a result of profligate consumption during the 1970s and 1980s. From a global perspective, it is clear that the privileged inhabitants of developed countries are acquiring cars, consuming petroleum and releasing CO_2 at a rate that far exceeds the global average. Yet, rather than seeking to moderate their consumption and transfer some of the resources to developing countries, industrialized nations are planning for further increases in personal mobility and resource consumption.

On top of the resources issue lie more recent concerns about the greenhouse effect. Global warming threatens our security on a global scale. No country, rich or poor, will be immune to the effects of a rapidly changing climate. Commitment and leadership will be needed from national governments, particularly in the developed world, in order to avert a climatic catastrophe. In the words of Chris Patten, formerly Secretary of State for the Environment in the British government:

> We must not allow uncertainty to inhibit us from taking action when the needs demand it. Precautionary action should be taken in the face of uncertainty when effects are irreversible or the risks of disaster too high.

References

Adams, J (1990) 'Car ownership forecasting: pull the ladder up, or climb back down?' *Traffic Engineering + Control*, March, pp 136–71

Alexander, W and Street, A (1989) *Metals in the Service of Man* ninth edition, Penguin, Harmondsworth

Armstrong, B (1983) *The influence of cool engines on car fuel consumption* Transport Research Laboratory, Crowthorne

Banister, C and D (1991) Personal communication

Barrett, S (1991) 'Global Warming: Economics of a Carbon Tax' in Pearce, D (ed) (1991) *Blueprint 2* Earthscan, London

Barrie, C (1990) 'VW "flask" keeps the heat on when the engine's off' *The Engineer* 5 April

Black, D (1992) 'Councils want M1 cash to update rail line' *The Independent* 11 January

Boehmer-Christiansen, S (1990) 'Vehicle Emission Regulation in Europe – the demise of lean-burn engines, the polluter pays principle... and the small car?' *Energy and Environment* Vol 1 No 1, pp 1–25

British Rail (1991) *InterCity File*

British Telecom (1991) *Energy, Telecommunications and the Environment*

Broom, S (1991) 'The lean burn engine' *Equinox* Simon Broom Associates, Channel 4

Cairncross, F (1991) *Costing the Earth* Business Books, London

Carr, G (1983) 'Potential for Aerodynamic Drag Reduction in Car Design' *International Journal of Vehicle Design*, Technological Advances in Vehicle Design Series, SP3, 'Impact of Aerodynamics on Vehicle Design', pp 44–56

Carter, P (1993) 'The Natural Advantage' *Coach and Bus Week* 10 July

Champion, A (1987a) 'Recent Changes in the Pace of Population Deconcentration in Britain' *Geoforum* Vol 18, No 4, pp 379–401

— (1987b) 'Momentous revival in London's population' *Town and Country Planning* Vol 56, No 3, pp 80–2

— (1990) Personal communication

Chang, T, Hammerle, R, Japar, S and Salmeen, I (1991) 'Alternative Transportation Fuels and Air Quality' *Environmental Science and Technology* Vol 25, No 7, pp 1190–7

Clover, C (1990) 'Patten urges car firms to stress mpg not mph' *Daily Telegraph* 22 September

Collier, C (1991) 'Trip Reduction Strategies: Will Commuters Change Travel Behaviour?' presented to the conference *Transportation and Global Climate Change: Long-Run Options* Pacific Grove, California, 25–8 August

Corrado, M (1990) 'Green Transport and the Consumer' presented to the conference *Environment Policy in Local Transport* London, 27 March

Cragg, C (1992) *Cleaning up Motor Car Pollution* Financial Times Business Information, London

Dallemagne, D (1990) *An Assessment of Potential Alternative Fuels for Vehicles in Europe* MSc dissertation, Imperial College Centre for Environmental Technology, London

Davies, C (1991) 'Environmental locomotives' *Modern Railways* January, p 45

Department of Energy (1984) *Energy Use and Energy Efficiency in UK Manufacturing Industry up to the year 2000, Volume 1* HMSO, London

— (1990) *An Evaluation of Energy-Related Greenhouse Gas Emissions and Measures to Ameliorate Them* Energy Paper 58, HMSO, London

Department of the Environment (1993) *Digest of Environmental Protection and Water Statistics* HMSO, London

Department of the Environment and Department of Transport (1993) *Reducing Transport Emissions Through Planning* HMSO, London

Department of Trade and Industry (1992) *Energy Related Carbon Emissions in Possible Future Scenarios for the United Kingdom* Energy Paper 59, HMSO, London

Department of Transport (1988) *National Travel Survey: 1985/86 Report – Part 1* HMSO, London

— (1989a) *National Road Traffic Forecasts (Great Britain)* HMSO, London

— (1989b) *Roads for Prosperity* HMSO, London

— (1990) *New Car Fuel Consumption – The Official Figures* HMSO, London

— (1991) 'Automatic vehicle speed data' *Road Accidents Great Britain 1990* HMSO, London

— (1992a) *Transport Statistics Great Britain* HMSO, London

— (1992b) *Tradeable credits to reduce CO_2 emissions from motor cars* HMSO, London

— (1992c) *Road Traffic Statistics Great Britain 1992* HMSO, London

— (1992d) *Assessing the Environmental Impact of Road Schemes* HMSO, London

— (1993a) *Vehicle Licensing Statistics 1992* HMSO, London

— (1993b) *Vehicle Speeds in Great Britain 1992* HMSO, London

— (1993c) *A Review of Technology for Road Use Pricing in London* HMSO, London

Dix, M and Goodwin, P (1982) 'Petrol prices and car use: a synthesis of conflicting evidence' *Transport Policy Decision Making* No 2, pp 179–95

Dobson, A, Jolly, A and Rubenstein, D (1989) 'The Greenhouse Effect and Biological Diversity' *Trends in Ecology and Evolution* Vol 4, No 3, pp 64–68

Dymock, E (1991) 'Technology transforms smoky old two-stroke' *The Sunday Times* 12 May

The Economist (1990) 'Japan's minicar market' 10 March

The European Commission (1992) *The Impact of Transport on the Environment: Sustainable Mobility* Brussels, February

European Federation for Transport and Environment (1992) *Making Fuel Go Further* EFTE, Brussels

Fergusson, M and Holman, C (1990) *Atmospheric Emissions from the Use of Transport in the United Kingdom Volume Two: The Effect of Alternative Transport Policies* WWF, England

Fisher, D, Hales, C, Wang, W-C, Ko, M and Sze, N (1990) 'Model calculations of the relative effects of CFCs and their replacements on global warming' *Nature* Vol 344, No 6266, pp 513–16

Fishlock, D (1991) 'Little engine goes to Detroit' *Financial Times* 19 April

Francis, R and Woollacott, P (1981) *Prospects for improved fuel economy and fuel flexibility in road vehicles* Energy Paper 45, HMSO, London

Gardner, D (1991) 'EC energy tax moves a stage closer' *Financial Times* 14 December

Gawthorpe, R (1983) 'Train drag reduction from simple design changes' *International Journal of Vehicle Design*, op cit (see Carr, 1983), pp 342–53

Giuliano, G and Wachs, M (1993) 'Employee Trip Reduction in Southern California: First Year Results' *Transportation Research* Vol 27A, No 2, pp 125–37

Gooding, K (1991) 'Quantum leap from cans to cars' *Financial Times* 21 June

Goodwin, P (1988) *Evidence on Car and Public Transport Demand Elasticities* Transport Studies Group, University of Oxford

— (1991) *Managing Traffic to Reduce Environmental Damage* Transport Studies Group, University of Oxford

— (1993) 'Confronting traffic growth: Ten final thoughts' in Stonham, P (ed) *Local Transport Today and Tomorrow*, Landor, London

Goodwin, P, Hallett, S, Kenny, F and Stokes, G (1991) *Transport: The New Realism* Transport Studies Unit, University of Oxford

Gotz, H (1983) 'Bus design features and their aerodynamic effects' *International Journal of Vehicle Design* op cit (see Carr, 1983), pp 229–55

Gould, R and Gribbin, J (1989) 'Greener cars may warm the world' *New Scientist* 20 May

Greene, D (1989) 'CAFE or PRICE?: An Analysis of the Effects of Federal Fuel Economy Regulations and Gasoline Price on New Car MPG, 1978–89' *The Energy Journal* Vol 11, No 3, pp 37–57

Gribbin, J (1989) 'The end of the ice ages?' *New Scientist* 17 June

Griffiths, J (1990) 'The cat takes on a new life' *Financial Times* 20 September

— (1991) 'Crystals help to cut cars' warm-up times' *Financial Times* 17 December

— (1992) 'Sparks fly over electric car' *The Independent* 24 March

Haines, A (1990) 'The Implications for Health', in Leggett, J (ed) op cit

Hall, D (1991) 'Altogether misguided and dangerous' *Town and Country Planning* Vol 60, No 11/12, pp 350–1

Hamer, M (1979) *Striking a Spark* Transport 2000, London

Hammond, A, Rodenburg, E and Moomaw, W (1991) 'Calculating National Accountability for Climate Change' *Environment* Vol 33, No 1, pp 10–35

Hass-Klau, C (1990) 'Greening the streets with public transport' *Urban Transport International* May/June

Heaps, C (1991) 'The Networker Revolution' in Network 2000, a *Rail* supplement

Highfield, R (1989) 'Global warming melts polar ice at British Antarctic base' *The Daily Telegraph*, 27 September

Hoffman, G and Zimdahl, W (1988) 'Recommended speed indication within the vehicle – a contribution to fuel saving through the "Wolfsburg Wave" information system' presented to the conference *Energy Efficiency in Land Transport* Commission of the European Communities, Luxembourg

Holman, C, Wade, J and Fergusson, M (1993) *Future Emissions from Cars 1990 to 2025: The Importance of the Cold Start Emissions Penalty* WWF, London

Homewood, B (1993) 'Will Brazil's cars go on the wagon?' *New Scientist* 9 January

Houghton, J, Jenkins, G and Ephraums, J (eds) (1990) *Climate Change: The IPCC Scientific Assessment* Cambridge University Press

House of Commons Energy Committee (1991) *Government Observations on the Third Report from the Committee (Session 1990–91) on Energy Efficiency* HMSO, London

Howard, D (1990) *Energy, Transport and the Environment*, Transnet, London

Hughes, P (1991) 'Exhausting the Atmosphere' *Town and Country Planning* Vol 60, No 9, pp 267–9

— (1992) *A strategy for reducing emissions of greenhouse gases from personal travel in the UK* PhD thesis, The Open University, Milton Keynes

— (1993) 'Road traffic on a plateau – but what happens next?' *Local Transport Today*, 30 September

Jackson, T and Schoon, N (1991) 'There's something electric in the air' *The Independent* 2 September

Knight, A and Cooke, J (1985) *Alternatives to Petrol* Standing Conference on Schools' Science and Technology, Hobsons, Cambridge

Lashof, D and Tirpak, D (1989) *Policy options for stabilizing global climate* USA Environmental Protection Agency

Laughren, F (1991) *Statement to the Legislature on Tax on Fuel Inefficient Vehicles*, 24 June, Ontario

Leggett, J (ed) (1990) *Global Warming: The Greenpeace Report*, Oxford University Press

— (1990) 'The Nature of the Greenhouse Threat', in Leggett, J (ed) (1990) op cit

Leone, R and Parkinson, T (1990) *Conserving Energy: Is There a Better Way? A study of corporate average fuel economy regulation* Boston University

Levenson, L and Gordon, D (1990) 'Drive+: promoting cleaner and more efficient motor vehicles through a self-financing system of state sales tax incentives' *Journal of Policy Analysis and Management* Vol 9, No 3, pp 409–15

Lewis, N (1993) *Road Pricing Theory and Practice* Thomas Telford, London

Lex Vehicle Leasing (1993) *1993 Lex Report on Motoring: The Company View*, Marlow

Local Transport Today (1992) 'York volunteers for European "car free city" club', 25 June

— (1993) 'THERMIE funding for energy-efficient transport in Liverpool and Hants' 8 July

London Evening Standard (1991) 'Scrap car tax', editorial, 21 February

Lucas Automotive (1990) *Delivering a Way of Life* Solihull

MacKenzie, J (1990) 'Ground Transportation: Implications for Global Warming' presented at the conference *The Route Ahead*, WWF, London

MacKinnon, I (1991) 'Londoners in favour of peak-time road fees' *The Independent* 27 September

Maltby, D, Monteath, I and Lawler, K (1978) 'The UK surface passenger transport sector: energy consumption and policy options for conservation' *Energy Policy* Vol 6, No 4, pp 294–313

Markandya, A (1991) 'Global Warming: The Economics of Tradeable Permits' in Pearce, D (ed) op cit

Martin, D and Michaelis, L (1992) *Road transport and the environment* Financial Times Business Information, London

Martin, D and Shock, R (1989) *Energy Use and Energy Efficiency in UK Transport Up to the Year 2010* Department of Energy Efficiency Series No 10, HMSO, London

McDiarmid, N (1992) 'British buses to run on flower power' *New Scientist* 3 October

McLain, L (1990) 'Little engine packs a punch' *Financial Times* 6 June

Metropolitan Transport Research Unit (MTRU) (1990) *Accessibility and Mobility: towards a useful measure of movement benefits* Discussion paper, London

— (1991) *Traffic Restraint: Five Cities, Five Solutions* London

Meyer, C (1993) 'Rough road ahead for biodiesel fuel' *New Scientist* 6 February

Meyers, S (1988) *Transportation in the LDCs: A Major Area of Growth in World Oil Demand* Lawrence Berkeley Laboratory, Berkeley

Millar, A (1993) 'UK Trials and Tribulations' *Coach and Bus Week* 10 July

Miller, D (1990) 'The market solution to reduce pollution' *The Independent* 29 July

Mills, E, Wilson, D and Johansson, T (1991) 'Getting started: no-regrets strategies for reducing greenhouse gas emissions' *Energy Policy* Vol 19, No 6, pp 526–42

Modern Railways (1986) 'Studies in Success: Bathgate reopening' June

Mogridge, M (1985) *Jam Yesterday, Jam Today and Jam Tomorrow?* University College London

Momenthy, A (1991) 'Characteristics of Future Aviation Fuels' presented to the conference *Transportation and Global Climate Change: Long-Run Options* Pacific Grove, California, 25–8 August

MORI (1990) *Public Attitudes Towards Transport and Pollution* WWF, London

National Audit Office (1989) *National Energy Efficiency* HMSO, London

Newman, P and Kenworthy, J (1989) *Cities and Automobile Dependence: An International Source Book* Gower, Aldershot

OECD (1982) *Automobile fuel consumption in actual traffic conditions*, OECD Road Research Group, Paris

— (1991) *OECD Environmental Data Compendium 1991* OECD, Paris

OPCS (1984) *1981 Census* HMSO, London

— (1990) OPCS Monitor, London

Owens, S (1986) *Energy, Planning and Urban Form* Pion, London

Patten, C (1989) Speech for Chatham House Conference, 5 December

Pearce, D, Markyanda, A and Barbier, E (1989) *Blueprint for a Green Economy* Earthscan, London

Pearce, F (1993) 'Carbon dioxide's taxing questions' *New Scientist* 26 June

Pearce T, and Waters, M (1980) *Cold start fuel consumption of a diesel and a petrol car*, Transport Research Laboratory, Crowthorne

Pendrous, R (1990) 'Hydraulic gear box may help car fuel economy' *The Engineer* 12 July

Pendyala, R, Goulias, K and Kitamura, R (1991) *Impact of Telecommuting on Spatial and Temporal Patterns of Household Travel: An Assessment for the State of California Pilot Project Participants* Institute of Transportation Studies, University of California, Davis

Potter, S (1990) 'Rail electrification: environmental and political influences on future trends', presented to the conference *Evolution of Electric Traction 1890-1990*

— (1991) 'The Impact of Company-Financed Motoring on Public Transport' 1991 Public Transport Symposium, University of Newcastle upon Tyne, 9–11 April

— (1992a) 'Integrating Fiscal and Transport Policies' in Roberts, J et al (eds) *Travel Sickness* Lawrence and Wishart, London

— (1992b) 'Funding and Investment in an Integrated Transport Policy' in Roberts J et al (eds) op cit

Potter, S and Hughes, P (1990) *Vital Travel Statistics*, Transport 2000, London

Radford, T (1992) 'Polar hole exposes climate's history' *The Guardian* 15 July

Ramanathan, V (1988) 'The Greenhouse Theory of Climate Change: A Test by an Inadvertent Global Experiment' *Science*, Vol 240, No 4850, pp 293–9

Redsell, M, Lucas, G and Ashford, N (1988) *Comparison of on-road fuel consumption for diesel and petrol cars*, Transport Research Laboratory, Crowthorne

Rowell, A and Fergusson, M (1991) *Bikes not Fumes* Cyclists' Touring Club, Godalming

Royal Town Planning Institute (1991) *Traffic Growth and Planning Policy* London

Saft Nife Ltd (1993) *Electric vehicles: Saft partners with US car makers* Press release, 4 March, Hampton

Schimel, D (1990) 'Biogeochemical Feedbacks in the Earth System', in Leggett (ed) (1990) op cit

Schipper, L (1991) 'Improved energy efficiency in the industrialized countries: past achievements, CO_2 emission prospects' *Energy Policy* Vol 19, No 2, pp 127–37

Schipper, S, Steiner, R and Santini, D (1991) 'Travel and energy use: the value of the international experience' presented to the conference *Transportation and Climate Change: Long-Run Options* Pacific Grove, California, 25–8 August

Schoon, N (1991) 'A far cry from the milk float' *The Independent* 2 September

Simpson, R (1993) 'Avoiding the particulate trap' *Coach and Bus Week* 10 July

Smith, R (1993) 'Ethanol, a Clean Future' *Coach and Bus Week*, 10 July

Society of Motor Manufacturers and Traders (SMMT) (1990) *The Motor Industry and the Greenhouse Effect*, London

South Coast Air Quality Management District (1989) *The Path to Clean Air: Attainment Strategies*, Los Angeles

Sperling, D and DeLuchi, M (1989) 'Transportation Energy Futures' *Annual Review of Energy* No 14, pp 375–424

Steadman, P and Barrett, M (1991) *The Potential Role of Town and Country Planning in Reducing Carbon Dioxide Emission* Centre for Configurational Studies, The Open University, Milton Keynes

Sturt, A (1992) 'Going Dutch' *Town and Country Planning* Vol 61, No 2, pp 48–51

TEST (1991) *Changed Travel – Better World?* London

Timberlake, L (1988) 'Global weather: what on Earth is happening?' *Telegraph Weekend*, 5 November

Transport Research Laboratory (1987) *Demonstration Safety Car – ESV87*, Transport Research Laboratory, Crowthorne

Viewpoint '89 (1989) 'Can Polar Bears Tread Water?' Channel 4

von Hippel, F and Levi, B (1983) 'Automobile Fuel Efficiency: The Opportunity and the Weakness of Existing Market Incentives' *Resources and Conservation* No 10, pp 103–24

Warren, A (1990) 'How Germany is rising to the carbon challenge' *The Guardian* 19 October

Watkins, L (1991) *Air pollution from road vehicles* Transport Research Laboratory, Crowthorne

Watson, R (1989) *Car fuel consumption: its relationship to official list consumptions* Transport Research Laboratory, Crowthorne

Watson, R, Rodhe H, Oeschger, H and Siegenthaler, U (1990) 'Greenhouse Gases and Aerosols', in Houghton et al (eds) (1990) op cit

Watts, S (1993) 'Computer group aims to keep work at a distance' *The Independent* 22 April

Webb, J (1991) 'Hydrogen-powered electric car sets sceptics wondering' *New Scientist* 29 June

Weiss, C (1990) 'Ethyl Alcohol as a Motor Fuel in Brazil' *Technology in Society* Vol 12, pp 255–82

Westbrook, F and Patterson, P (1989) 'Changing Driving Patterns and their Effect on Fuel Economy' presented at the SAE Government/Industry meeting, 2 May, Washington DC

White, P (1986) *Comparison of Vehicle Costs* Paper C220/86, I Mech E, London

Whitelegg, J (1990a) 'European Instruments for an Environmental Transport Policy in the European Context' European Parliament hearing on *Economic and Fiscal Incentives to Promote Environmental Policy Objectives* Brussels, 21–2 June

— (1990b) 'The principle of environmental traffic management' in Tolley, R (ed) (1990) *The Greening of Urban Transport* Belhaven Press, London

— (1991) 'ETR – a step beyond the fiscal quick fix' *Town and Country Planning* Vol 60, No 9, pp 270–3

— (1993) *Transport for a Sustainable Future: The Case for Europe* Belhaven Press, London

Withrington, P (1990) Personal communication

Woodwell, G (1990) 'The Effects of Global Warming', in Leggett (ed) (1990) op cit

Wootton, J (1993) 'Local transport solutions with 2020 vision' in Stonham, P (ed) *Local Transport Today and Tomorrow* Landor, London

World Meteorological Organisation (1991) *Scientific Assessment of Ozone Depletion: 1991* WMO Ozone Report No 25

Zeevenhooven, N (1990), 'The energy consumption of various transport systems' in *Railways, Environment and Transport Quality*, a collection of expert papers prepared for the International Transport Workers' Federation, Utrecht

Index

Index compiled by John Tooke.